Monographs on Endocrinology

Volume 8

Edited by

F. Gross, Heidelberg · A. Labhart, Zürich · M. B. Lipsett, Cleveland
T. Mann, Cambridge · L. T. Samuels, Salt Lake City
J. Zander, München

E. Gurpide

Tracer Methods in Hormone Research

With 35 Figures

Springer-Verlag Berlin · Heidelberg · New York 1975

Erlio Gurpide, Ph. D.

Professor of Biochemistry and of Obstetrics and Gynecology
Mount Sinai School of Medicine, City University of New York,
New York, NY 10029/USA

ISBN 978-3-642-80894-4 ISBN 978-3-642-80892-0 (eBook)
DOI 10.1007/978-3-642-80892-0

Library of Congress Cataloging in Publication Data. Gurpide, Erlio, 1927— Tracer methods in hormone research. (Monographs on endocrinology; v. 8). Bibliography: p. Includes index. 1. Hormones. 2. Tracers (Biology). 3. Compartmental analysis (Biology). I. Title. [DNLM: 1. Hormones. 2. Radioisotopes. W1 M057 v. 8 / WK102 G981t] QP571.G87 591.1'927 74-23519

Softcover reprint of the hardcover 1st edition 1975

Acknowledgments

This monograph is based mainly on topics developed in collaboration with Dr. JONAH MANN, Department of Mathematics, The City College, City University of New York, and published during the period 1963–1970.

It also includes discussions on "turn-over times" published with Dr. ALDO RESCIGNO, Department of Physiology, University of Minnesota.

The sections on numerical methods in Chapters 3 and 5 were written by R. J. BOGUMIL, Department of Obstetrics and Gynecology, Mount Sinai School of Medicine, New York. His help in annotating the text is also gratefully acknowledged.

Discussions with Drs. Paul C. MacDonald, Raymond L. Vande Wiele and Seymour Lieberman at Columbia University, New York, held during the development of tracer methods to determine rates of secretion of dehydroisoandrosterone sulfate, are pleasantly recalled. That project motivated theoretical treatments presented in this monograph.

Contents

Introduction

The purpose of this monograph is to describe theoretical aspects of the interpretation of data obtained from experiments performed with labeled hormones.

Quantitative endocrinologic studies involving the use of tracers include the determination of rates at which hormones are secreted by endocrine glands and are produced outside these glands by conversion of other secreted hormones. Tracer experiments are also performed with the purpose of measuring rates of metabolic reactions. These measurements reveal the contribution of secreted hormones to the formation of circulating compounds and urinary metabolites. The estimation of rates of fetal and placental production and exchange of hormones characterizes a class of *in vivo* quantitative studies performed with isotopically labeled hormones (radioactive or not). In addition, tracers are used to measure permeability and rates of reaction in *in vitro* systems, and to study the uptake of hormones by tissues, both *in vivo* and *in vitro*.

The stability of the steroid nucleus carrying the isotopic label and the large number of reversible metabolic reactions in which steroids are involved, both facilitated and motivated the development of a sophisticated theoretical treatment of tracer experiments in the field of endocrinology. Although the practical examples used to illustrate the concepts and calculations presented in this monograph involve labeled hormones, the theory is presented in a general symbolic manner and is applicable to other fields of investigation.

Different problems require the application of special experimental designs and lead to calculations that may not be found in a collection of formulas. Therefore, the intention of this monograph is to emphasize the description of procedures followed to derive presently available formulas and the assumptions on which these derivations are based, both to facilitate their proper application and to stimulate further advances. Formulas to calculate kinetic parameters from isotopic data represent logical consequences of *a priori* assumptions concerning the biologic system under study. The current assumptions may eventually become unsatisfactory and may be superseded by more realistic and sophisticated schemes.

Most of the analyses presented in this monograph refer to open systems in the steady state. Under these conditions, rates can be assigned constant values and clear relationships can be established between rates and isotope concentrations obtained in tracer experiments. In reality, however, many hormones are produced at a variable rate and steady state conditions are not commonly fulfilled. It is then important to note that the metabolic fate of a labeled hormone is usually (but not always) unaffected by changes in hormone production. Therefore, parameters with values estimated entirely from isotope measurements may be validly discussed in terms of steady state conditions, although the calculation of rates requires consideration of the variable rate of production of the hormones. These considerations are discussed throughout the monograph and particularly in Chapter 5.

The reader should not confuse mathematical difficulty, which is minimal in this monograph, with annoyances arising from the symbolism employed. Regardless of efforts to simplify the notation, the large number of parameters and experimental data to be handled makes the symbolism quite cumbersome. This is the most irritating feature of all publications on analysis of tracer experiments. In fact, sometimes it is easier to derive a formula than to decipher someone else's notation. This seems to be a common problem, as revealed by the high incidence of rediscovery of theorems in compartmental analysis.

The monograph is organized as follows:

Chapter 1 describes open systems in the steady state and provides definitions for rates of transfer between different pools or compartments is such systems. Various other parameters describing the dynamics of production and movement of material in the system are also defined, and the relationships among them are derived. No reference to tracers is made in this chapter. However, rates and parameters are defined with the advanced knowledge that they can be estimated from tracer experiments. In fact, the proper definition of rates of transfer was found to be a key point in the interpretation of isotopic data.

Chapter 2 provides the formulas necessary for the estimation of the parameters described in Chapter 1 using data obtained from experiments in which the tracers are infused at a constant rate until a steady state with respect to the isotopes is achieved. The derivation of these formulas is simply based on equations describing the conservation of labeled and unlabeled material in any space, at the steady state. The solution of the resulting systems of linear equations is quite elementary. This chapter emphasizes the advantages and possibilities offered by the simultaneous use of two tracers labeled with different isotopes.

Chapter 3 presents an analysis of experiments in which tracers are administered as a single dose rather than infused at a constant rate. It shows a theoretical relationship that exists between the isotopic data obtained in the same system by these two methods of administration of tracers. This simple relationship justifies the application of the formulas derived in Chapter 2 to data obtained after rapid injection of a tracer. Therefore, no differential equations are needed to derive the formulas used in the calculation of rates. This chapter also covers the procedures used in the determination of specific activity functions from isotopic data and describes parameters that are calculable when these functions are known.

In Chapter 4, a more conventional compartmental analysis is presented. Models consisting of several compartments are used to derive expressions describing the isotope content in each compartment in terms of the parameters of the system and the time following the administration of tracers. This is a formal mathematical analysis based on abstract models.

Chapter 5 deals with the interpretation of tracer concentration curves in biologic systems, with a specific reference to blood data. In contrast to the analysis in Chapter 4, the physiologic meaning of pools or compartments is questioned and the assumptions necessary to justify the calculations of rates are critically reviewed. This chapter includes a discussion of several "turnover" terms defined in Chapter 3 and describes applications of 2 isotope methods to the study of peripheral interconversion and fetomaternal transfer of hormones.

Chapter 6 examines the information made available by experiments in which the isotope content of urinary metabolites is determined after the administration of tracers. The interpretation of urinary data is complex because of the multiplicity of precursors of the metabolites, but deserves special attention inasmuch as urine is a convenient and widely used source of experimental data.

Chapter 7 describes an *in vitro* experimental design based on superfusion of tissue slices or cell suspensions with a mixture of metabolically related tracers. It shows how isotopic steady state data can be used to estimate rates of entry, exit, synthesis, and metabolism, as well as intracellular compartmentalization of compounds.

Although some of the formulas presented in this monograph can be derived without involving the concept of compartments, most of the theory is based on the consideration of multicompartmental models. In order to achieve physiologic and biochemical relevance through these models, a form of analysis was developed that is based on data obtained from a few accessible compartments among the many that exchange the isotope introduced into the system. The calculation of rates of movement of material into, out of, and between these selected pools can be validated even when the number and identity of the other pools in the network is unknown. Consequently, the complexity of the system need not be neglected by limitations imposed by the models used to interpret the isotopic data.

A fundamental characteristic of the systems considered in this monograph is that the movement of the labeled molecules out of a compartment can be described by a first-order expression, viz. that the rate of exit of the isotope towards any destination is always proportional to the amount of isotope in the compartment. A simple manner to assure this property is to specify that the system is at the steady state with respect to the unlabeled molecules, as is discussed in Note 3.2, Chapter 3. Because hormonal steady state may be the exception rather than the rule, it is important to realize that the movement of isotope may follow first-order kinetics even in systems that are not at the steady state, as is apparent in situations in which the kinetic behavior of the labeled molecules is not influenced by changes in the levels of the endogenous hormones. These systems can be studied with the tracer methodology described in this monograph. However, the proposed methods are not applicable to systems in which the values of the rate constants of processes involving the tracer vary during the experiment. These situations are apparently rare in the field of endocrinology, at least at the present level of experimental precision. Therefore, it can be expected that the concepts based on "first-order" tracer kinetics or even on steady state will continue to yield useful information on the production, distribution, and metabolism of hormones.

Chapter 1

Rates in Open System in the Steady State: Definitions and Relations

Labeled compounds introduced into living systems are distributed in various spaces and undergo chemical reactions. In order to give a structure to the system, it is usually assumed that there are regions in which a compound is uniformly distributed or, at least, in which internal dispersion occurs much more rapidly than does outward flow; a compound distributed in one such space constitutes a pool. The term "pool" is used in preference to "compartment" to avoid the purely spatial implication that might be assigned to the latter term.

In a biologic system the total number and the anatomic boundaries of most of the pools defined in this manner cannot be specified. Therefore, the definitions of rates and related parameters described in this monograph are chosen for their applicability to multicompartmental systems in which the total number of pools is unknown. When a rate of transfer between two pools is defined and a formula for its measurement from isotopic data is described, it is understood that the two pools are well defined and accessible to the investigator, even though they may be embedded in a complex network of many other pools, some also accessible, some hopelessly inaccessible. The definition of rates, as given here, allows for the existence of unknown intermediate pools in the path of movement of material between two pools. In contrast, typical compartmental analysis is based on models consisting of a determined and explicit number of pools.

The manner in which rates are defined in this chapter is a key to the development of simple tracer equations applicable to systems of an unknown number of pools (HART, 1966; MANN and GURPIDE, 1966).

Formulas for the calculation of each of these rates from experimental data are given in subsequent chapters.

One, two, three, or, in general, m accessible pools embedded in a network of an unknown number of compartments may be considered in any particular study, depending upon the number of pools that are sampled or directly labeled. These cases are discussed following the order of increasing complexity. Definitions of average rates in non-steady-state systems are discussed in other chapters.

A. One Pool Embedded in a Multicompartmental System

Rates of Entry

Figure 1.1 represents one of the pools in the system, viz. pool 1. This pool may represent a hormone in systemic circulation. The molecules of the compound entering the pool can be divided into two categories: molecules entering the pool for the first time (at a rate v_{01}) and those which are reentering the pool (at a rate w_1). The sum of these two rates represents the total amount of material coming into the pool per unit of time.

Production Rate

The symbol PR_1 denotes the "production rate" of the compound in the pool, i.e., the rate at which *new* material is fed into the pool. It contrasts with the "unproductive" recycle of material denoted by w_1. When only one pool embedded in a multicompartmental system is considered, PR_1 equals v_{01}. Originally the term production rate was used to distinguish the total rate of *de novo* formation of a hormone from the "secretion rate" of the hormone by endocrine glands (VANDE WIELE *et al.*, 1963), because some hormones are produced by more than one gland as well as by metabolic conversion of other secreted hormones. This parameter is easily determined when a tracer is administered into the pool (Eq. 2.1, Chapter 2).

Rates of Removal

The compound leaves the pool in one of two ways: irreversibly, at a rate v_{10}, or temporarily, bound to return to the pool. The latter rate is, by definition, equal to w_1. At the steady state, PR_1 equals v_{10}, since the rate at which the compound enters *de novo* into the pool must equal the rate at which it irreversibly leaves the pool.

The processes represented by the rates v_{10} and w_1 do not necessarily refer to different metabolic paths. Rather, each path of removal of the compound from the pool may contribute to both of these two rates. Formulas to calculate w_1 from isotopic data are given in Chapter 3.

B. Two Related Pools Embedded in a Multicompartmental System

Rates of Entry, Exit, and Transfer

Figure 1.2 represents two pools between which there is a transfer of material in at least one direction. Pools 1 and 2 may represent two blood-borne hormones such as estrone and estradiol between which interconversion is noted,

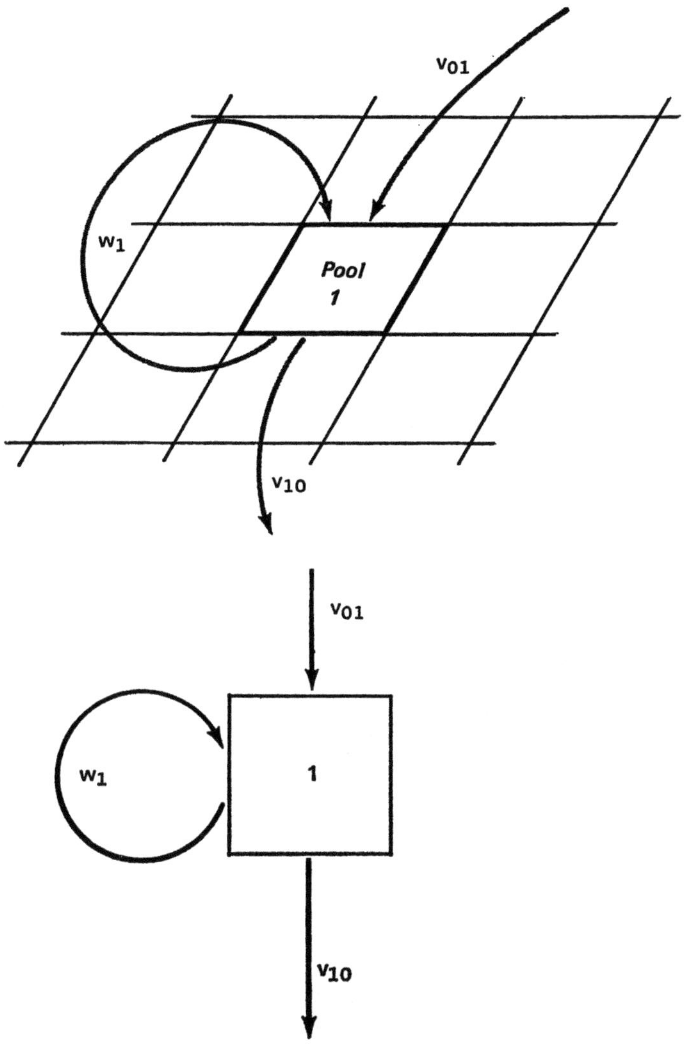

Fig. 1.1. Representation of a pool embedded in a multicompartmental system. Pool 1 represents a compound uniformly distributed in a space in which mixing occurs instantaneously or, in experimental terms, in which internal dispersion occurs much more rapidly than outward flow. In a biologic system, a pool exchanges material with an undetermined number of other pools represented by the network on the upper figure. Material appears *de novo* in pool 1 at a rate v_{01} and leaves the pool irreversibly at a rate v_{10}. Some material leaving the pool returns to it a rate w_1. The lower figure is the simplest representation of the system; the cycle indicated by the arrow implies the existence of other pools exchanging material with pool 1

or they may designate one hormone distributed in two spaces, such as cortisol in fetal and maternal circulations.

The rate at which material is transferred from pool 1 to pool 2 is denoted by v_{12}. It describes the total transfer, by all paths, of material from one pool

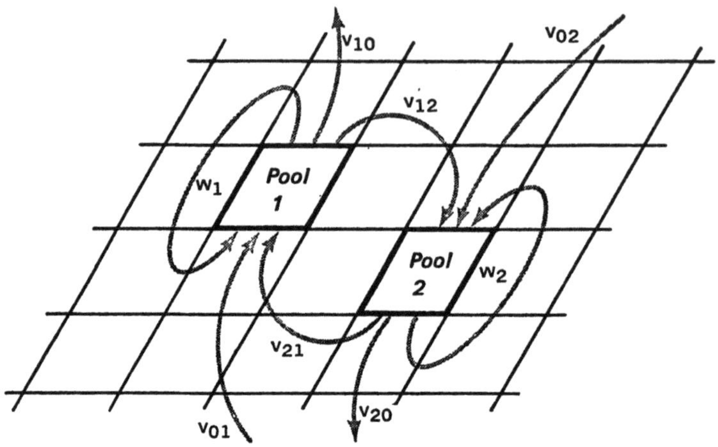

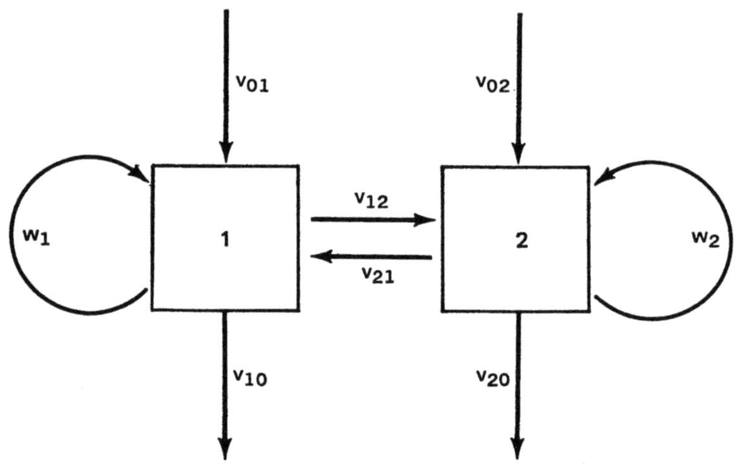

Fig. 1.2. Representation of two pools embedded in a multicompartmental system. Another pool of the network (pool 2) has been singled out for examination of its relation to pool 1 (see Fig. 1.1). Note that the transfer of material between pools 1 and 2 (rates v_{12} and v_{21}) may occur through other intermediate pools. Note also that v_{01} and w_1 in Fig. 1.2 may be smaller than v_{01} and w_1 in Fig. 1.1. Superscripts for these symbols (e.g., $v_{01}^{(2)}$) are used in the text only when the possibility of confusion arises. The lower figure is the simplest representation of this system; the cycle arrows imply the existence of an undetermined number of exchanging pools

to the other. The rate of transfer from pool 2 to pool 1 is v_{21}; the net exchange between the two pools is expressed by the difference between these two rates.

Each rate of transfer, e.g., v_{12}, can be considered to be the sum of two other rates:

$$v_{12} = v_{12}' + v_{12}'' \tag{1.1}$$

where v_{12}' refers to material that has never been in pool 2, and v_{12}'' refers to material that is returning to pool 2. Such a distinction is justified because both

v'_{12} and v''_{12} can be measured by tracer experiments and because v'_{12} is a component of the production rate of the compound in pool 2.

All material entering pool 1 may be classified in one of three categories: (1) new material not coming from pool 2 (rate v_{01}), (2) recycled material not coming from pool 2 (rate w_1), or (3) material coming from pool 2 (rate v_{21}).

All material leaving pool 1 may be similarly divided into three categories according to whether it is (1) irreversibly removed without going to pool 2 (v_{10}); (2) returned to pool 1 by any path except through pool 2 (w_1); or (3) transferred to pool 2 (v_{12}). All of these rates can be estimated from isotopic data after administration of tracers into pools 1 and 2, as shown in other chapters.

Similar definitions can be used to describe the rates at which material enters and leaves pool 2.

The production rate of pool 1 (PR_1) was defined in the preceding section as the rate of entry into pool 1 of material that had not previously been in that pool, regardless of its origins. Consideration of a second pool (pool 2 in Fig. 1.2) serves to describe further the origin of new material entering pool 1. According to the definition of production rate,

$$PR_1 = v_{01} + v'_{21} \tag{1.2}$$

where v'_{21} is that part of v_{21} concerning material that has never been in pool 1. Thus, it is possible to determine from tracer data the sources of the compound in pool 1.

In contrast to the production rate, other rate values depend on the number of pools under consideration. The value of w_1, for example, will vary depending on whether or not pool 2 is being considered. Thus, w_1 in Fig. 1.1 is larger, by definition, than w_1 in Fig. 1.2, even though pool 1 is the same in both cases. Similarly, v_{10} in Fig. 1.1 is larger than v_{10} in Fig. 1.2, because part of the total rate of irreversible removal of material from pool 1 is included in the rate v_{21}. Therefore, it may be necessary occasionally to indicate with a superscript attached to symbols for rates, the number of pools in the system studied. Thus, $v_{10}^{(1)}$ and $v_{10}^{(2)}$ denote the values of the rates of irreversible removal of material from the system through pool 1 when either one or two pools are considered. No superscripts are used unless they are required for clarity. The symbol for production rate, a pool parameter that is independent of the number of pools considered, requires no superscript.

The Transfer Factor

The transfer factor of material moving from pool 1 to pool 2, ϱ_{12}, is defined by the expression

$$\varrho_{12} = \frac{v'_{12}}{v_{01}}. \tag{1.3}$$

This expression states that ϱ_{12} is the fraction of new material entering pool 1 from all sources other than pool 2 that eventually reaches pool 2.

Similarly, $\varrho_{21} = v'_{21}/v_{02}$. An important relation follows from this equation and from Eq. 1.2:

$$PR_1 = v_{01} + \varrho_{21} v_{02}. \tag{1.4}$$

The meaning of ϱ_{12} is further clarified by expressions equivalent to Eq. 1.3 obtained from the following considerations.

It can be intuitively seen that, in the steady state, the correspondence between v'_{12} and v_{01} in Eq. 1.3 also must hold for v''_{12} and v_{21}, i.e.,

$$\varrho_{12} = \frac{v''_{12}}{v_{21}}. \tag{1.5}$$

According to this expression, ϱ_{12} represents the fraction of the rate of material transferred from pool 2 to pool 1 that returns to pool 2. The proof of the equivalence between Eq. 1.3 and 1.5 is given in Chapter 2.

Furthermore, because simple algebra indicates that $r = (a+c)/(b+d)$ if $r = a/b = c/d$, it follows from Eqs. 1.3 and 1.5 that

$$\varrho_{12} = \frac{v'_{12}}{v_{01}} = \frac{v''_{12}}{v_{21}} = \frac{v'_{12} + v''_{12}}{v_{01} + v_{21}} = \frac{v_{12}}{v_{01} + v_{21}}$$

or, at the steady state,

$$\varrho_{12} = \frac{v_{12}}{v_{01} + v_{21}} = \frac{v_{12}}{v_{10} + v_{12}}. \tag{1.6}$$

The ratio of the rates v_{12} and v_{10} is then given by the expression

$$\frac{v_{12}}{v_{10}} = \frac{\varrho_{12}}{1 - \varrho_{12}}. \tag{1.7}$$

A question now arises as to the meaning of the product $\varrho_{12} PR_1$. The answer can be obtained formally by making $\varrho_{12} = x/PR_1$ and replacing PR_1 by the expression in Eq. 1.4. These substitutions yield the solution $x = \varrho_{12} v_{01} + \varrho_{12} \varrho_{21} v_{02}$, or

$$\varrho_{12} = \frac{v'_{12} + \varrho_{12} v'_{21}}{PR_1}. \tag{1.8}$$

Therefore, ϱ_{12} also represents the fraction of PR_1 that travels the path from pool 1 to pool 2 for the first time. In summary,

$$\varrho_{12} = \frac{v_{12}}{v_{12} + v_{10}} = \frac{v_{12}}{v_{01} + v_{21}} = \frac{v'_{12}}{v_{01}} = \frac{v''_{12}}{v_{21}} = \frac{v'_{10} + \varrho_{12} v'_{21}}{PR_1}.$$

It is apparent from Eq. 1.6 that ϱ_{12} has the following properties:

$$\begin{aligned}
&\varrho_{12} \geqq 0 && \text{since } v_{12} \geqq 0 \text{ and } v_{01} + v_{21} > 0 \\
&\varrho_{12} \leqq 1 && \text{since } v_{10} \geqq 0 \\
&\varrho_{12} = 0 && \text{if } v_{12} = 0 \\
&\varrho_{12} = 1 && \text{if } v_{10} = 0.
\end{aligned}$$

The emphasis placed upon the meaning of the transfer factor ϱ is justified because this parameter can be easily determined in tracer experiments (see Chapter 2) and can be used in conjunction with values of production rates to estimate rates, as described in the next section.

Calculations of Rates from Values of Production Rates (PR's) and Transfer Factors (ϱ's)

Among the parameters that are calculated most easily in tracer experiments are the PR's and the ϱ's (Chapter 2). Average values of PR's of most circulating hormones in normal human subjects are known and ϱ values for the conversion of one blood-borne hormone to another have been determined for a number of hormonal pairs (BAIRD *et al.*, 1968, 1969) (see Tables 5.1 and 5.2). If PR_1, PR_2, ϱ_{12}, and ϱ_{21} are known, all the v's in Fig. 1.2 can be calculated. The pertinent formulas are derived as follows.

As indicated by Eq. 1.4,

$$PR_1 = v_{01} + \varrho_{21} v_{02}$$
$$PR_2 = \varrho_{12} v_{01} + v_{02} .$$

This system of two independent linear equations with two unknowns (v_{01} and v_{02}) can be solved by elementary methods, e.g., by determinants using Cramer's rule (see, e.g., NAHIKIAN, 1964). Thus,

$$v_{01} = \frac{PR_1 - \varrho_{21} PR_2}{1 - \varrho_{12} \varrho_{21}} \tag{1.9}$$

and

$$v_{02} = \frac{PR_2 - \varrho_{12} PR_1}{1 - \varrho_{12} \varrho_{21}} . \tag{1.10}$$

Another system of equations may be used to calculate v_{12} and v_{21}. From Eq. 1.6,

$$\varrho_{12} = \frac{v_{12}}{v_{01} + v_{21}} \qquad \text{and} \qquad \varrho_{21} = \frac{v_{21}}{v_{02} + v_{12}}$$

or

$$v_{12} - \varrho_{12} v_{21} = \varrho_{12} v_{01}$$
$$-\varrho_{12} v_{12} + v_{21} = \varrho_{21} v_{02} .$$

Replacing v_{01} and v_{02} by their values in terms of PR's and ϱ's (Eqs. 1.9 and 1.10), it follows that

$$v_{12} - \varrho_{12} v_{21} = \varrho_{12} \frac{PR_1 - \varrho_{21} PR_2}{1 - \varrho_{12} \varrho_{21}}$$
$$-\varrho_{12} v_{12} + v_{21} = \varrho_{21} \frac{PR_2 - \varrho_{12} PR_1}{1 - \varrho_{12} \varrho_{21}} .$$

This pair of equations in the unknowns v_{12} and v_{21} yields the solution

$$v_{12} = \frac{\varrho_{12} PR_1}{1 - \varrho_{12} \varrho_{21}} \tag{1.11}$$

and

$$v_{21} = \frac{\varrho_{21} PR_2}{1 - \varrho_{12} \varrho_{21}} . \tag{1.12}$$

Another procedure to derive Eq. 1.11 is of conceptual interest. The rate v_{12} can be considered to equal the sum of the rates at which material is transferred from pool 1 to pool 2 for the first time ($\varrho_{12}\,PR_1$), for the second time ($\varrho_{12}\,PR_1 \times \varrho_{21}\,\varrho_{12}$), for the third time ($\varrho_{12}\,PR_1 \times \varrho_{21}\,\varrho_{12} \times \varrho_{21}\,\varrho_{12}$), and so on. Then,

$$v_{12} = \varrho_{12}\,PR_1\,[1 + \varrho_{21}\,\varrho_{12} + (\varrho_{21}\,\varrho_{12})^2 + \cdots]$$

$$\text{or } v_{12} = \frac{\varrho_{12}\,PR_1}{1 - \varrho_{12}\,\varrho_{21}}$$

$$\text{since } \frac{1}{1 - \varrho_{12}\,\varrho_{21}} = 1 + \varrho_{21}\,\varrho_{12} + (\varrho_{21}\,\varrho_{12})^2 + \cdots .$$

Finally, it follows from the steady-state conditions that

$$v_{10} = v_{01} + v_{21} - v_{12} = PR_1\,(1 - \varrho_{12}) / (1 - \varrho_{12}\,\varrho_{21}) \tag{1.13}$$

and

$$v_{20} = v_{02} + v_{12} - v_{21} = PR_2\,(1 - \varrho_{21}) / (1 - \varrho_{12}\,\varrho_{21}) . \tag{1.14}$$

These results are included in Table 1.1 in which all parameters are expressed in terms of PR's and ϱ's.

The rates w_1 in Fig. 1.1, and w_1 and w_2 in Fig. 1.2, can also be calculated from isotopic data (Chapter 3). The data necessary to calculate these rates are, however, different from the data used to determine PR's, ϱ's, and all the parameters listed in Table 1.1.

Further Description of the Rate of Transfer of Material from One Pool to Another

It follows from the definition of the transfer factor ϱ_{12} that

$$v_{12} = \varrho_{12}\,(v_{01} + v_{21})$$

or

$$v_{12} = \varrho_{12}\,v_{01} + \varrho_{12}\,v'_{21} + \varrho_{12}\,v''_{21} .$$

The meaning of each of these terms, illustrated in Fig. 1.3, is as follows $\varrho_{12}\,v_{01}$ represents the rate of transfer of material that has never been in pool 2 ($\varrho_{12}\,v_{01} = v'_{12}$, Eq. 1.3); $\varrho_{12}\,v'_{21}$ represents the rate of transfer of material that already has been in pool 2 and returns to pool 1 for the first time; and

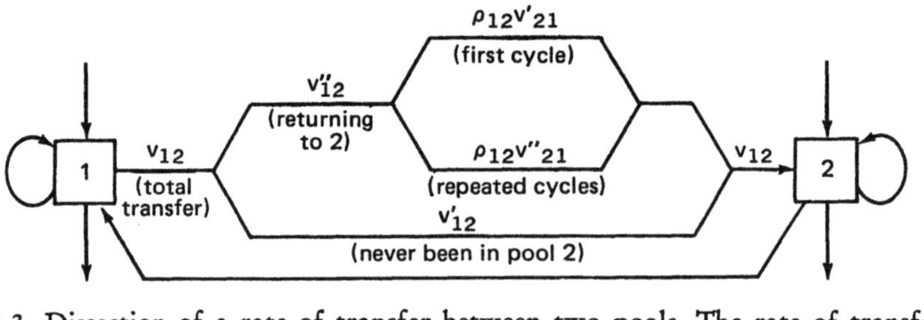

Fig. 1.3. Dissection of a rate of transfer between two pools. The rate of transfer of material from pool 1 to pool 2 (v_{12}) is the sum of the rates of transfer of material that has never been in pool 2 (v'_{12}) or is returning to pool 2 (v''_{12}) for the first time ($\varrho_{12}\,v'_{21}$) or by repeated cycles ($\varrho_{12}\,v''_{21}$). The transfer factor ϱ_{12} equals the ratio $v_{12}/(v_{12}+v_{10})$. These various rates can be calculated from the same isotopic data necessary to estimate v_{12} (see Chapter 2)

Table 1.1. Formulas used to calculate parameters of entry, transfer, and removal in two related pools embedded in a multiple pool system (Fig. 1.2) [a]

Parameter	Formula
Rate of entry into pool 1 from outside (v_{01})	$v_{01}=\dfrac{PR_1-\varrho_{21}PR_2}{1-\varrho_{12}\varrho_{21}}$
Rate of entry into pool 2 from outside (v_{02})	$v_{02}=\dfrac{PR_2-\varrho_{12}PR_1}{1-\varrho_{12}\varrho_{21}}$
Contribution factors	
Of 1 to 2 (Δ_{12})	$\Delta_{12}=\varrho_{12}\dfrac{PR_1}{PR_2}$
Of 2 to 1 (Δ_{21})	$\Delta_{21}=\varrho_{21}\dfrac{PR_2}{PR_1}$
Rate of transfer from 1 to 2	
Total (v_{12})	$v_{12}=\dfrac{\varrho_{12}PR_1}{1-\varrho_{12}\varrho_{21}}$
"Never before in pool 2" (v'_{12})	$v'_{12}=(1-\Delta_{21})\,v_{12}$
Recycled (v''_{12})	$v''_{12}=\Delta_{21}\,v_{12}$
First cycle	$\varrho_{12}\,v'_{21}$
More than one cycle	$\varrho_{12}\,v''_{21}$
Rate of transfer from 2 to 1	
Total (v_{21})	$v_{21}=\dfrac{\varrho_{21}PR_2}{1-\varrho_{12}\varrho_{21}}$
"Never before in pool 1" (v'_{21})	$v'_{21}=(1-\Delta_{12})\,v_{21}$
Recycled (v''_{21})	$v''_{21}=\Delta_{12}\,v_{21}$
First cycle	$\varrho_{21}\,v'_{12}$
More than one cycle	$\varrho_{21}\,v''_{12}$
Rate of irreversible removal	
Out of pool 1 (v_{10})	$v_{10}=PR_1\dfrac{1-\varrho_{12}}{1-\varrho_{12}\varrho_{21}}$
Out of pool 2 (v_{20})	$v_{20}=PR_2\dfrac{1-\varrho_{21}}{1-\varrho_{12}\varrho_{21}}$

[a] From data on production rates (PR's) and transfer factors (ϱ').

$\varrho_{12}\,v''_{21}$ includes material that has already gone through at least one cycle between the two pools.

The introduction of components of the total rate of transfer is not arbitrary. These rates, as well as all other parameters defined in this chapter, can be estimated from isotopic data. In fact, the same data necessary for the calculation of v_{12} allow the estimation of the three component rates (see Chapter 2).

The Contribution Factor

A basic question that arises during the analysis of two-pool systems is: how much of the material in one pool derives from the other? To avoid ambiguity, this question must be precisely expressed in terms of rates of transfer and removal. For this purpose, the definition of a "contribution factor", readily measurable in tracer experiments, is introduced. The contribution of pool 1 to pool 2, Δ_{12}, is defined as

$$\Delta_{12} = \frac{v_{12}}{v_{02} + v_{12}}. \tag{1.15}$$

Therefore, the relative contribution of the rates v_{12} and v_{02} to the formation of pool 2 is given by the expression

$$\frac{v_{12}}{v_{02}} = \frac{\Delta_{12}}{1 - \Delta_{12}}. \tag{1.16}$$

Intuitively, if Δ_{12} represents the fraction of pool 2 made by material received from pool 1, then v''_{21} should equal $\Delta_{12}\, v_{21}$, i.e.,

$$\Delta_{12} = \frac{v''_{21}}{v_{21}}. \tag{1.17}$$

A formal justification of the identity of Eqs. 1.15 and 1.17 is presented in Chapter 2 where it is shown that the isotopic data necessary to calculate v_{21} allow the estimation of Δ_{21}, and therefore of v''_{21} and v'_{21}.

The relation between Δ_{12} and PR_2 can be derived as follows. Replace v_{12} in the numerator of Eq. 1.15 by the expression obtained from the definition of ϱ_{12} (Eq. 1.6). Hence, $\Delta_{12}/\varrho_{12} = (v_{01} + v_{21})/(v_{02} + v_{12})$. Since $v_{01} + v_{21} = v_{01} + v'_{21} + v''_{21} = PR_1 + v''_{21}$, then $\Delta_{12}/\varrho_{12} = (PR_1 + v''_{21})/(PR_2 + v''_{12})$. From Eqs. 1.17 and 1.5, it results that

$$\Delta_{12} = \frac{\varrho_{12}\, v''_{21}}{v''_{12}}.$$

Then,

$$\frac{\Delta_{12}}{\varrho_{12}} = \frac{v''_{21}}{v''_{12}} = \frac{PR_1 + v''_{21}}{PR_2 + v''_{12}}.$$

Since, in general, $r = (a - c)/(b - d)$, if $r = a/b = c/d$, it follows that

$$\frac{\Delta_{12}}{\varrho_{12}} = \frac{PR_1}{PR_2}$$

or

$$\Delta_{12} = \frac{\varrho_{12}\, PR_1}{PR_2}. \tag{1.18}$$

It is also of interest to determine the relation that exists between the factor Δ_{12} and the ratios v'_{12}/v_{02} or v'_{12}/PR_2. On the basis of Eqs. 1.3, 1.9, 1.10, and 1.18, the following algebraic relations are obtained:

$$\frac{v'_{12}}{v_{02}} = \frac{\varrho_{12}\, v_{01}}{v_{02}} = \varrho_{12}\, \frac{PR_1 - \varrho_{21}\, PR_2}{PR_2 - \varrho_{12}\, PR_1} = \frac{\Delta_{12} - \varrho_{12}\, \varrho_{21}}{1 - \Delta_{12}} = \frac{\Delta_{12}(1 - \Delta_{21})}{1 - \Delta_{12}} \tag{1.19}$$

and

$$\frac{v'_{12}}{PR_2} = \frac{v'_{12}}{v'_{12} + v_{02}} = \frac{1}{1 + (v_{02}/v'_{12})} = \frac{\Delta_{12}(1 - \Delta_{21})}{1 - \Delta_{12}\, \Delta_{21}} \leq \Delta_{12}. \tag{1.20}$$

In summary,

$$\Delta_{12} = \frac{v_{12}}{v_{12}+v_{02}} = \frac{v_{12}}{v_{21}+v_{20}} = \frac{v''_{21}}{v_{21}} = \varrho_{12}\,\frac{PR_1}{PR_2}\,.$$

A contribution factor Δ_{01} can be similarly defined as

$$\Delta_{01} = \frac{v_{01}}{v_{01}+v_{21}}\,. \tag{1.21}$$

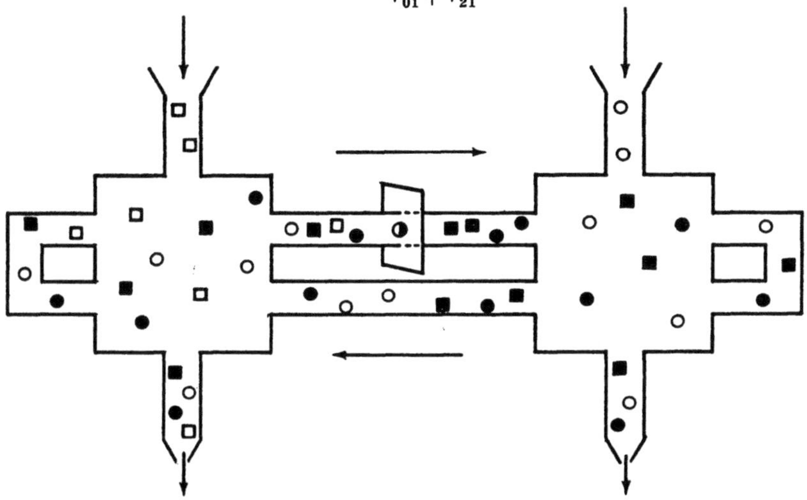

Fig. 1.4. Distinction between material in a pool according to origin. This model represents two pools embedded in a multicompartmental system. Pool 1 (on the left side) receives new material from outside the system from pools other than pool 2 (on the right of the figure). This new material is represented by white squares. New material may come to pool 1 from pool 2 in the form of white circles. Material transferred from pool 1 to 2 for the first time is assumed to be tagged during the transfer: white squares or white circles returning to pool 2, are filled (as illustrated by the changes in the color of a circle crossing a hypothetical plane that intercepts the path between pools 1 and 2). Note that recirculation of material around pool 1 through other pools not involving pool 2 does not alter the pattern of shapes and colors of the elements in pool 1. From the concepts given in the text, it follows that the steady-state composition of pool 1 can be predicted from the values of contribution factors as indicated in the list given below. Formulas for the estimation of Δ's from isotopic data are are given in Chapter 2.
Pool 1:

$$\frac{\text{White squares}}{\text{All elements}} = \frac{v_{01}}{v_{01}+v_{21}} = \Delta_{01} = 1 - \Delta_{21}$$

$$\frac{\text{White circles}}{\text{All elements}} = \frac{v'_{21}}{v_{01}+v_{21}} = \Delta_{02}\,\Delta_{21}$$

$$\frac{\text{White elements}}{\text{All elements}} = \frac{PR_1}{v_{01}+v_{21}} = \Delta_{01} + \Delta_{02}\,\Delta_{21} = 1 - \Delta_{12}\,\Delta_{21}$$

$$\frac{\text{White circles}}{\text{White squares}} = \frac{\text{All circles}}{\text{All squares}} = \frac{v'_{21}}{v_{01}} = \frac{\Delta_{02}\,\Delta_{21}}{\Delta_{01}}$$

$$\frac{\text{White circles}}{\text{White elements}} = \frac{\text{All circles}}{\text{All elements}} = \frac{v'_{21}}{PR_1} = \frac{\Delta_{21}\,\Delta_{02}}{1 - \Delta_{12}\,\Delta_{21}}$$

$$\frac{\text{White squares}}{\text{White elements}} = \frac{\text{All squares}}{\text{All elements}} = \frac{v_{01}}{PR_1} = \frac{\Delta_{01}}{1 - \Delta_{12}\,\Delta_{21}}\,.$$

Therefore,

$$\Delta_{01} + \Delta_{21} = 1 \, .$$

Production rates can be related to the Δ's by expression derived from Eq. 1.4, i.e.,

$$PR_1 = v_{10} + \Delta_{12} \, v_{20}$$

and

$$PR_1 = v_{01} + \Delta_{02} \, v_{21} \, .$$

The contribution factors serve to describe the portion of a pool made by material originating from a specific source. This concept is emphasized in Fig. 1.4 and in the explanatory legend to this figure.

Application of the Two-Pool Analysis to a Hormonal System

The application of the formulas given in this section (Table 1.1) to a particular hormonal system can be illustrated using values of production rates and transfer factors of circulating estrone and estradiol in normal young men,

Table 1.2. Rates (nmole/day) and contribution factors (%) characterizing the exchange among circulating estrone (E_1), estradiol (E_2), and estrone sulfate (E_1S) in normal men

Range		E_1-E_2; E_1-E_1S; E_2-E_1S					$E_1-E_2-E_1S$				
i	j	$\Delta_{ij}^{(2)}$ (×100)	$v_{ij}^{(2)}$	$v_{ij}^{\prime(2)}$	$v_{ij}^{\prime\prime(2)}$	$\varrho_{ij}\, v_{ji}^{\prime(2)}$	$\Delta_{ij}^{(3)}$ (×100)	$v_{ij}^{(3)}$	$v_{ij}^{\prime(3)}$	$v_{ij}^{\prime\prime(3)}$	$v_{ij}^{\prime\prime\prime(3)}$
E_1	E_2	7.7	13	11.2	1.8	1.6	6.7	11.4	9.8	1.6	1.1
E_2	E_1	13.7	35.0	32.3	2.7	2.3	9.7	26.5	24.5	2.0	1.8
E_1	E_1S	42.0	109	93.4	15.6	9.0	40	105	90	15	14.7
E_1S	E_1	14.3	38.9	22.6	16.3	14.0	14	38.3	22.2	16.1	15.3
E_2	E_1S	28.7	70.8	68.5	2.3	1.6	23	61	59	2	1.3
E_1S	E_2	3.2	5.4	3.8	1.6	1.5	2.2	3.8	2.7	1.1	0.9

[a] Rates are calculated from the experimental values listed in the legend to Fig. 1.7.

[b] Valves and symbols shown in the first row are described by the following narrative. A total of 13 nmole of circulating E_1 is converted to circulating E_2 per day ($v_{E_1E_2}^{(2)}$). Most of the material transferred ($v_{E_1E_2}^{\prime(2)}$) has never been in circulation as E_2 and only 1.8 nmole/day corresponds to recycling material ($v_{E_1E_2}^{\prime\prime(2)}$) of which 1.6 nmole/day ($\varrho_{E_1E_2}\, v_{E_1E_2}^{\prime(2)}$) is in the first cycle. The rate of conversion of E_1 to E_2 without the intermediacy of blood-borne E_1S is 11.4 nmole/day ($v_{E_1E_2}^{(3)}$). Of this rate, 9.8 nmole/day corresponds to material that has never been plasma E_2 before ($v_{E_1E_2}^{\prime(3)}$), whereas 1.6 nmole/day ($v_{E_1E_2}^{\prime\prime(3)}$) corresponds to recycle. In this rate of recycle, a part ($v_{E_1E_2}^{\prime\prime(3)} = 1.1$) corresponds to recycle without intermediacy of E_1S in either direction. A 7.7% of E_2 in circulation derives from blood-borne E_1 ($\Delta_{E_1E_2}^{(2)}$), but only 6.7% is derived from E_1 converted to E_2 without the intermediate formation of E_1S ($\Delta_{E_1E_2}^{(3)}$).

published by LONGCOPE *et al.* (LONGCOPE *et al.*, 1968; LONGCOPE and TAIT, 1971). The data and results are shown in Fig. 1.5 and in Table 1.2. Table 1.2 gives further details on the structure of the rates of interconversion, and also includes the evaluation of the extent of participation of estrone sulfate in the exchange between estrone and estradiol according to procedures described in the next section.

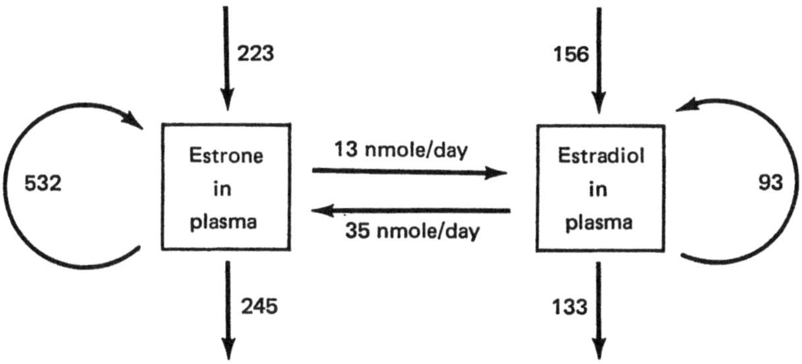

Fig. 1.5. Rates (nmole/day) of formation and metabolism of blood-borne estrone and estradiol in normal young men. The values used to calculate the rates shown in Fig. 1.5 are the following:

$\varrho_{E_1E_2} = 0.05$ (LONGCOPE *et al.*, 1968)

$\varrho_{E_2E_1} = 0.21$ (LONGCOPE and TAIT, 1971)

$MCR_{E_1} = 1170$ liter plasma/(day×m²) (LONGCOPE and TAIT, 1971).

If an average surface area of 1.96 m² is considered (LONGCOPE *et al.*, 1968), then

$MCR_{E_1} = 2300$ liter plasma/day

$MCR_{E_2} = 910$ liter plasma/(day×m²) (LONGCOPE and TAIT, 1971)
or 1780 liter plasma/day

$c_{E_1} = 30.1$ ng/liter plasma $= 0.111$ nmole/liter (NAGAI and LONGCOPE, 1971)

$c_{E_2} = 25.6$ ng/liter plasma $= 0.094$ nmole/liter (NAGAI and LONGCOPE, 1971).

Consequently,

$PR_{E_1} = MCR_{E_1}\, c_{E_1} = 256$ nmole/day

$PR_{E_2} = MCR_{E_2}\, c_{E_2} = 167$ nmole/day.

The formulas used to calculate the rates shown in Fig. 1.5 are listed in Table 1.1. Rates of recycle (w_1 and w_2 in Fig. 1.2) were estimated as described in Chapter 3 (Eq. 3.27) and in Note 5.2, Chapter 5, using data on disappearance of labeled E_1 and E_2 from plasma after a single intravenous injection of ^3H-E_1 and ^3H-E_2 (LONGCOPE and WILLIAMS, 1974), kindly made available to us by Dr. C. LONGCOPE before publication. Although these data correspond to a nonpregnant woman, the calculated w values are included in Fig. 1.5 to illustrate the extent of recycle of the hormones. Figures 1.5 and 1.7 can be compared to evaluate the participation of circulating estrone sulfate in the interconversion of E_1 and E_2

C. Three Related Pools Embedded in a Multicompartmental System (Fig. 1.6)

Rates of Entry, Exit, and Transfer

Figure 1.6 represents three pools among which there is transfer of material. As in the previous cases in which only one or two pools were considered, it is understood that the biologic system is more complex, involving a large number of communicating pools. The three pools selected for examination are part of such a large system.

The total rate of entry of material into pool 1 can be divided in the following distinct rates according to its origin:

v_{21}	From pool 2, bypassing pool 3
v_{31}	From pool 3, bypassing pool 2
w_1	From pool 1 bypassing pools 2 and 3
v_{01}	From any other source.

The total rate of removal from pool 1 ($v_{T,1}$) also falls into four groups according to the fate of the material leaving the pool:

v_{12}	Transfer to pool 2, bypassing pool 3
v_{13}	Transfer to pool 3, bypassing pool 2
v_{10}	Irreversible removal from the system, bypassing pools 2 and 3
w_1	Reentry into pool 1, bypassing both pools 2 and 3.

Production Rates

The production rate of one of the three pools under consideration can be related to the other two pools by the following expression which is justified in the next chapter (Eq. 2.38):

$$PR_1 = v_{01} + \varrho_{21} v_{02} + \varrho_{31} v_{03} \tag{1.22}$$

where PR_1 is a characteristic value of pool 1, which does not change with the number of other pools included in the system. The ϱ's are defined by Eq. 1.6, in terms of two pools, and no other definition for transfer factors is introduced throughout this monograph. In contrast, v_{01} changes in value according to the number of pools under consideration.

Equation 1.22 is intuitively apparent since the rate at which material enters pool 1, *de novo* (PR_1), equals the sum of rates of entry of material into the three-pool system, each multiplied by the fraction transferred "for the first time" from the pool of entry to pool 1, regardless of the pathway of transfer.

Each of the rates of transfer between pools can be resolved into two components. For instance, $v_{13} = v'_{13} + v''_{13}$ where v'_{13} denotes the rate at which material that has not been previously in pool 3 is transferred from pool 1 to pool 3,

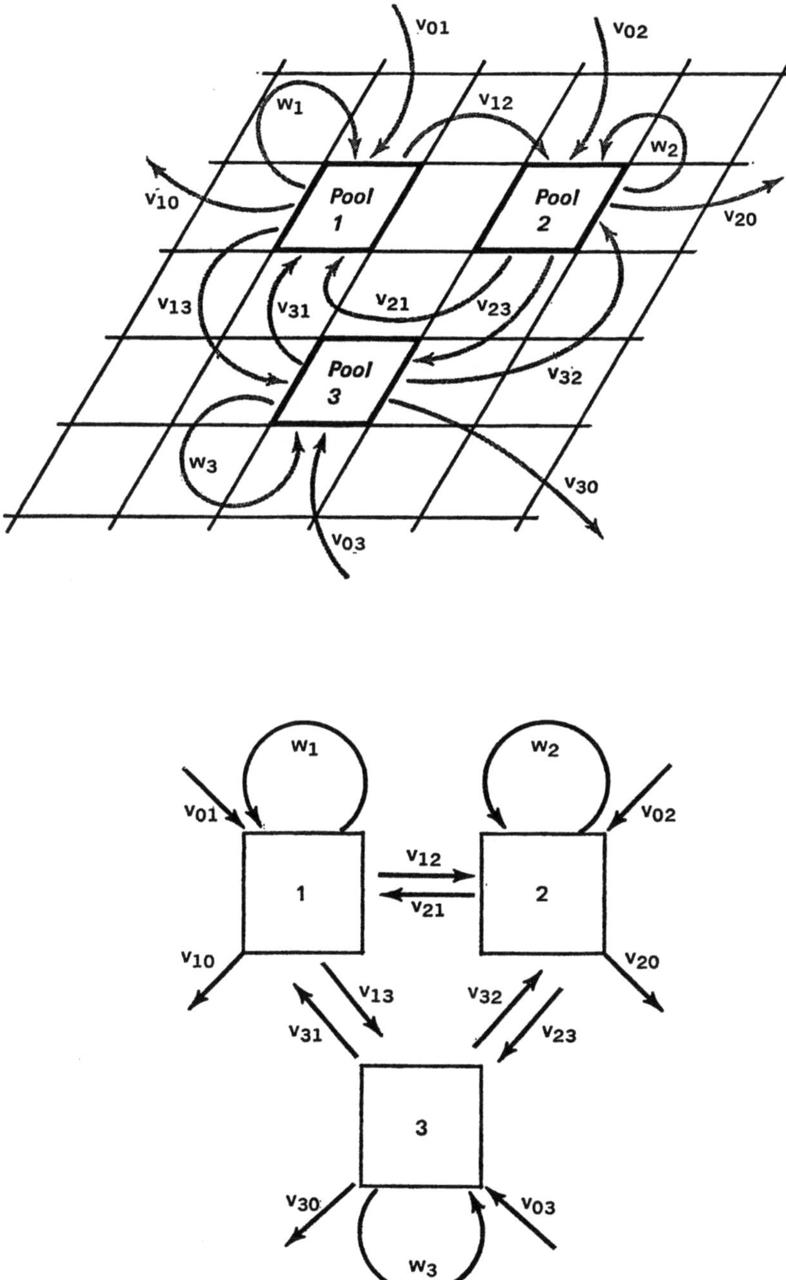

Fig. 1.6. Representation of three pools embedded in a multicompartmental system. Three of the undetermined number of exchanging pools in the system are singled out for study. Note that v_{13} represents the rate of transfer of material from pool 1 to pool 3 by all paths except those passing through pool 2. Also note that v_{01}, v_{10}, v_{12}, and w_1 may be smaller than the corresponding rates in Fig. 1.3. The lower figure is the simplest representation of this system, with the w arrows indicating the presence of other exchanging pools

bypassing pool 2. Therefore, v''_{13} represents the rate at which material returns to pool 3 from pool 1, bypassing pool 2.

According to these definitions, it follows that

$$PR_1 = v_{01} + v'_{21} + v'_{31}. \tag{1.23}$$

It is intuitively apparent by analogy with Eq. 1.17, that in a three-pool subsystem

$$v''^{(3)}_{21} = \Delta^{(2)}_{12} v^{(3)}_{21}. \tag{1.24}$$

The superscripts in parentheses indicate the number of pools in the subsystem under consideration. $\Delta^{(2)}_{12}$ denotes the fraction of pool 2 made by material transported from pool 1 to 2 regardless of the path by which that transfer occurs. Therefore, $\Delta^{(2)}_{12} v^{(3)}_{21}$ is the rate at which material returns from pool 1 to pool 2 without the intermediacy of pool 3. Consequently,

$$v'^{(3)}_{21} = v^{(3)}_{21} - v''^{(3)}_{21} = v^{(3)}_{21} (1 - \Delta^{(2)}_{12}). \tag{1.25}$$

In general,

$$v''^{(3)}_{ij} = \Delta^{(2)}_{ji} v^{(3)}_{ij} \qquad i, j = 1, 2, 3, \quad i \neq j \tag{1.26}$$

and

$$v'^{(3)}_{ij} = (1 - \Delta^{(2)}_{ji}) v^{(3)}_{ij} \qquad i, j = 1, 2, 3, \quad i \neq j. \tag{1.27}$$

Contribution Factor

A contribution factor in a three-pool system is defined in a manner similar to that for a two-pool system. For instance, the factor $\Delta^{(3)}_{13}$, which described the contribution to pool 3 of material transferred from pool 1 bypassing pool 2, is defined by the expression

$$\Delta^{(3)}_{13} = \frac{v_{13}}{v_{03} + v_{13} + v_{23}} = \frac{v_{13}}{v_{30} + v_{31} + v_{32}}. \tag{1.28}$$

An equation analogous to Eq. 1.24 can be written to calculate the rate (v'''_{21}) at which material that went from 1 to 2 without passing through pool 3 is returning to pool 2 without the intermediacy of pool 3, i.e.,

$$v'''^{(3)}_{21} = \Delta^{(3)}_{12} v^{(3)}_{21}. \tag{1.29}$$

In general,

$$v'''^{(3)}_{ij} = \Delta^{(3)}_{ji} v^{(3)}_{ij} \qquad i, j = 1, 2, 3, \quad i \neq j. \tag{1.30}$$

Calculation of Rates from Values of PR's and ϱ's

Rates of entry, transfer, and exit can be calculated if the values of production rates for each pool and transfer factors between all pairs of pools are known.

It follows from the analysis in Chapter 2 (Eq. 2.30) that

$$
\begin{aligned}
-\frac{1}{PR_1}(v_{01}+v_{21}+v_{31})+\frac{\varrho_{12}}{PR_2}v_{21}+\frac{\varrho_{13}}{PR_3}v_{31}&=-1\\
-\frac{\varrho_{21}}{PR_1}(v_{01}+v_{21}+v_{31})+\frac{1}{PR_2}v_{21}+\frac{\varrho_{23}}{PR_3}v_{31}&=0\\
-\frac{\varrho_{31}}{PR_1}(v_{01}+v_{21}+v_{31})+\frac{\varrho_{32}}{PR_2}v_{21}+\frac{1}{PR_3}v_{31}&=0\,.
\end{aligned}
\qquad (1.31)
$$

This system of equations can be solved by determinants using Cramer's rule; expressions for v_{01}, v_{21}, and v_{31} in terms of PR's and ϱ's are then obtained.

Similar systems of equations corresponding to transfers to pools 2 and 3 can be written. Therefore, v_{02}, v_{03} and all rates of transfer and removal can be calculated. Consequently, the Δ's can also be obtained in terms of PR's and ϱ's. The resulting formulas are shown in Table 1.3.

Table 1.3. Formulas used to calculate rates and contribution factors in a three-pool subsystem (Fig. 1.6) from values of production rates (PR's) of the 3 pools and transfer factors (ϱ's) among them [a]

$$\text{Denominator}=D=1+\varrho_{12}\varrho_{23}\varrho_{31}+\varrho_{13}\varrho_{32}\varrho_{21}-\varrho_{12}\varrho_{21}-\varrho_{13}\varrho_{31}-\varrho_{23}\varrho_{32}$$

$$
\begin{aligned}
v_{01}&=[PR_1(1-\varrho_{23}\varrho_{32})-PR_2(\varrho_{21}-\varrho_{23}\varrho_{31})-PR_3(\varrho_{31}-\varrho_{32}\varrho_{21})]/D\\
v_{02}&=[PR_2(1-\varrho_{13}\varrho_{31})-PR_1(\varrho_{12}-\varrho_{13}\varrho_{32})-PR_3(\varrho_{32}-\varrho_{12}\varrho_{31})]/D\\
v_{03}&=[PR_3(1-\varrho_{12}\varrho_{21})-PR_1(\varrho_{13}-\varrho_{12}\varrho_{23})-PR_2(\varrho_{23}-\varrho_{21}\varrho_{13})]/D
\end{aligned}
$$

$$
\begin{aligned}
v_{12}&=PR_1(\varrho_{12}-\varrho_{13}\varrho_{32})/D & \Delta_{12}&=PR_1(\varrho_{12}-\varrho_{13}\varrho_{32})/PR_2(1-\varrho_{13}\varrho_{31})\\
v_{21}&=PR_2(\varrho_{21}-\varrho_{23}\varrho_{31})/D & \Delta_{21}&=PR_2(\varrho_{21}-\varrho_{23}\varrho_{31})/PR_1(1-\varrho_{23}\varrho_{32})\\
v_{13}&=PR_1(\varrho_{13}-\varrho_{12}\varrho_{23})/D & \Delta_{13}&=PR_1(\varrho_{13}-\varrho_{12}\varrho_{23})/PR_3(1-\varrho_{12}\varrho_{21})\\
v_{31}&=PR_3(\varrho_{31}-\varrho_{32}\varrho_{21})/D & \Delta_{31}&=PR_3(\varrho_{31}-\varrho_{32}\varrho_{21})/PR_1(1-\varrho_{23}\varrho_{32})\\
v_{23}&=PR_2(\varrho_{23}-\varrho_{21}\varrho_{13})/D & \Delta_{23}&=PR_2(\varrho_{23}-\varrho_{21}\varrho_{13})/PR_3(1-\varrho_{12}\varrho_{21})\\
v_{32}&=PR_3(\varrho_{32}-\varrho_{31}\varrho_{12})/D & \Delta_{32}&=PR_3(\varrho_{32}-\varrho_{31}\varrho_{12})/PR_2(1-\varrho_{13}\varrho_{31})
\end{aligned}
$$

$$v_{10}=v_{01}+v_{21}+v_{31}-v_{12}-v_{13} \qquad v_{20}=v_{02}+v_{12}+v_{32}-v_{21}-v_{23}$$

$$v_{30}=v_{03}+v_{13}+v_{23}-v_{31}-v_{32}$$

[a] Formulas were obtained by solving the system of Eqs. 1.31.

Application of the Three-Pool Analysis to a Hormonal System

Values for the production rate of circulating estrone sulfate and transfer factors corresponding to the interconversions between circulating estrone sulfate (E_1S), estrone (E_1), and estradiol (E_2) in men are available from reports by LONGCOPE (1972) and RUDER *et al.* (1972). These experimental values can be combined with those corresponding to E_1 and E_2 to obtain the results shown in Fig. 1.7. These results were obtained by using the formulas shown in Table 1.3, which are based on PR and ϱ values. The dissection of the rates of transfer according to their return components is done according to Eqs. 1.26, 1.27, and 1.30. The results are shown in Table 1.2 (Note 1.1).

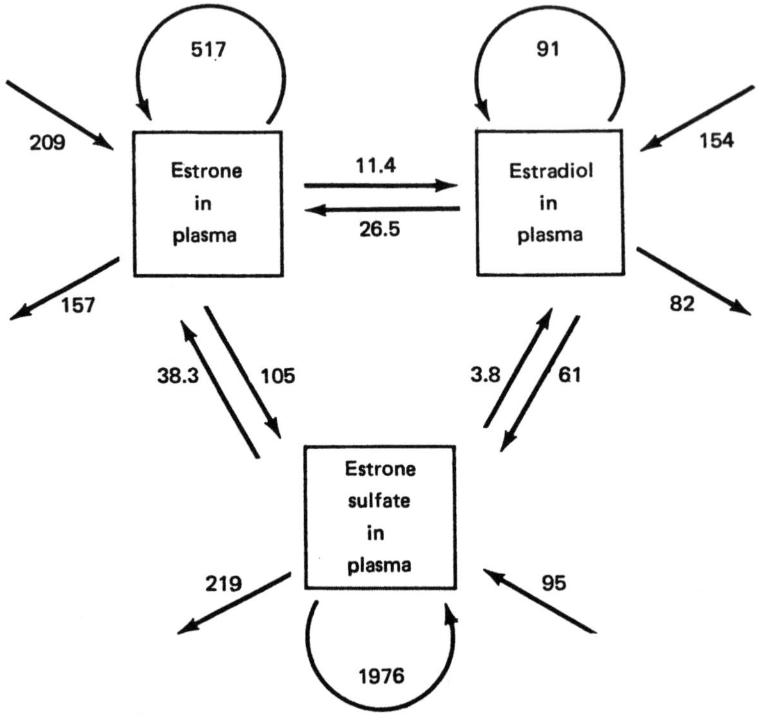

Fig. 1.7. Rates (nmole/day) of formation and metabolism of blood-borne estrone and estrone sulfate in normal young men. The following values were used to calculate the rates shown in Fig. 1.7. As indicated in the legend to Fig. 1.5,

$PR_{E_1} = 256$ nmole/day, $\varrho_{E_1E_2} = 0.05$, $c_{E_1} = 0.111$ nmole/liter plasma

$PR_{E_2} = 167$ nmole/day, $\varrho_{E_2E_1} = 0.21$, $c_{E_2} = 0.094$ nmole/liter plasma

$MCR_{E_1S} = 200$ liter plasma/day

$\varrho_{E_1S,E_1} = 0.15$, $\varrho_{E_1E_1S} = 0.40$

$\varrho_{E_1S,E_2} = 0.022$, $\varrho_{E_2E_1S} = 0.42$ (LONGCOPE, 1972)

$c_{E_1S} = 330$ pg E_1 as E_1S/ml plasma $= 1.22$ nM (LORIAUX *et al.*, 1971).

Consequently,

$PR_{E_1S} = 244$ nmole/day.

The rates shown in Fig. 1.7 were calculated using the formulas shown in Table 1.3, obtained by solving the system of Eq. 1.31. Table 1.2 shows a more detailed analysis of the conversion rates using the experimental data given above. Rates of recycle (w_1, w_2 and w_3 in Fig. 1.6) were estimated from data on the decline of the levels of tracer in plasma following the intravenous injection of E_1S (LONGCOPE, 1972), E_1 and E_2 (LONGCOPE and WILLIAMS, 1974) and the concentrations of E_1, E_2, and E_1S in plasma listed above (see Note 5.3 and legend to Fig. 1.5). Data on MCR's and ϱ's corresponding to the system E_1-E_2-E_1S have also been reported by RUDER *et al.* (1972) (Note 1.1)

Note 1.1

The example in Fig. 1.7 is used to describe how to calculate rates from experimental data rather than to discuss specific problems related to estrogen dynamics. It is of interest to note, however, that the results obtained with LONGCOPE's data indicate a significant production of E_1S from sources other than circulating E_1 and E_2. In contrast, the experimental data presented by RUDER *et al.* (1972), although not very different from those reported by LONGCOPE (1972) and LONGCOPE and TAIT (1968), yield negative values for the production of E_1S not peripherally derived from E_1 and E_2. As RUDER *et al.* point out, imprecision in these calculations is a necessary consequence of using values obtained from different subjects. The normal range of variation of MCR, c, PR, and ϱ values among individuals is large. Therefore, data from two consecutive experiments (e.g., $^3H\text{-}E_2 + {}^{14}C\text{-}E_1$ followed by $^{14}C\text{-}E_1 + {}^3H\text{-}E_1S$) performed on a single subject may be more valuable in the evaluation of independent production of E_1S than average PR and ϱ values.

D. General Systems of m Related Pools

The concept of the existence of discrete spaces in which the distribution of a compound is faster than the rates at which the compound is removed from the space can be extrapolated to define pools in a mathematical sense, by specifying instantaneous and homogeneous mixing within the pool. An entire biochemical system may then be represented as a system of mathematically defined pools, if no restrictions on the number of pools are imposed.

Some pools may be unrelated in the sense that no material is transferred between them. Others may be related unidirectionally, i.e., the flow of material from one pool to the other is irreversible.

In an "irreducible" ("strongly connected", RESCIGNO and SEGRE, 1966) system, all pools exchange material with each other directly or indirectly. If a tracer were injected into any pool belonging to an irreducible system, all the pools in the system would be labeled.

Direct transfer of material from one pool to another indicates that such transfer does not occur through an intermediate pool. Transfer from one pool to another can be direct, indirect or both, since there may be several paths of transport between pools.

In the approach used in previous sections, m of a total number n of theoretical pools in the system, are selected for study. These are called primary pools (MANN and GURPIDE, 1966). The advantage of this approach is that any calculations involving only these m pools recognizes the existence of the other pools and therefore does not constitute an arbitrary oversimplification of the actual system. Furthermore, because m may be made theoretically as large as n, no generality is lost by using this form of analysis.

Some expressions given in previous sections of this chapter can be generalized to m primary pool systems. For this purpose, the subindices i, j, k, etc., will refer to different pools in the system. The pools are assumed to be numbered from 1 to m.

A general definition for steady-state production rate of material in a pool would then be

$$PR_i = v_{0i}^{(m)} + \sum_{\substack{j=1 \\ j \neq i}}^{m} v_{ji}^{(m)} \qquad i = 1, 2, \ldots, m . \tag{1.31}$$

It can be intuitively seen, and is formally proved in the next chapter (Eq. 2.38), that

$$PR_1 = v_{0i}^{(m)} + \sum_{\substack{j=1 \\ j \neq i}}^{m} \varrho_{ji} \, v_{0i}^{(m)} \qquad i = 1, 2, \ldots, m \tag{1.32}$$

where the ϱ factors are defined in terms of two primary pools, i.e.,

$$\varrho_{ji} = \frac{v_{ji}^{(2)}}{v_{0j}^{(2)} + v_{ji}^{(2)}} \qquad i, j = 1, 2, \ldots, m, \qquad i \neq j .$$

The values of the rates $v_{0i}^{(m)}$ can be determined from experimentally measured PR's and ϱ's by solving the system of linear algebraic equations 1.32.

The symbol $-v_{ii}^{(m)}$ may be defined as the sum of all the rates at which material leaves pool i with the exception of the rate of recirculation $w_i^{(m)}$, i.e.,

$$-v_{ii}^{(m)} = \sum_{\substack{j=0 \\ j \neq i}}^{m} v_{ij}^{(m)} \qquad i = 1, 2, \ldots, m . \tag{1.33}$$

In a system at the steady state, $-v_{ii}^{(m)}$ equals the sum of the rates of entry of material into the pool from outside the system and from the primary pools, i.e.,

$$-v_{ii}^{(m)} = \sum_{\substack{j=0 \\ j \neq i}}^{m} v_{ji}^{(m)} \qquad i = 1, 2, \ldots, m . \tag{1.34}$$

Equation 1.34 can then be more succinctly written as

$$\sum_{j=0}^{m} v_{ji}^{(m)} = 0 \qquad i = 1, 2, \ldots, m . \tag{1.35}$$

The contribution factors may now be defined in general as

$$\Delta_{ij}^{(m)} = \frac{v_{ij}^{(m)}}{-v_{jj}^{(m)}} \qquad i, j = 1, 2, \ldots, m, \qquad i \neq j . \tag{1.36}$$

The contribution to pool j of material entering the pool from outside the system can be denoted by the factor $\Delta_{0j}^{(m)}$, which is defined as follows:

$$\Delta_{0j}^{(m)} = \frac{v_{0j}^{(m)}}{-v_{jj}^{(m)}} \qquad i = 1, 2, \ldots, m . \tag{1.37}$$

From these definitions, it follows that

$$\sum_{\substack{i=0 \\ i \neq j}}^{m} \Delta_{ij}^{(m)} = 1 \qquad j = 1, 2, \ldots, m. \tag{1.38}$$

The various contribution factors of pool i to pool j indicate the fractions of pool j derived from material transferred from pool i by all paths ($\Delta_{ij}^{(2)}$) or without the intermediacy of other pools in the system, $\Delta_{ij}^{(m)}$. Therefore,

$$v_{ji}^{\prime\prime(m)} = \Delta_{ij}^{(2)}\, v_{ji}^{(m)} \qquad i, j = 1, 2, \ldots, m, \qquad i \neq j \tag{1.39}$$

indicates that part of the total rate $v_{ji}^{(m)}$ which represents material returning to pool i. Furthermore,

$$v_{ji}^{\prime\prime\prime(m)} = \Delta_{ij}^{(m)}\, v_{ji}^{(m)} \qquad i, j = 1, 2, \ldots, m, \qquad i \neq j \tag{1.40}$$

indicates the rate at which material originally transferred from i to j without passing through the other primary pools returns from pool j to pool i by the same path.

A transfer factor, always defined on the basis of two primary pools, can be described in terms of rates in an m pool system as follows (Gurpide *et al.*, 1963):

$$\varrho_{ij} = \frac{v_{ij}^{(2)}}{-v_{ii}^{(2)}} = \frac{C_{ij}}{C_{jj}} \qquad i, j = 1, 2, \ldots, m, \qquad i \neq j \tag{1.41}$$

where C_{ij} is the cofactor of the element in the i-th row and the j-th column in the determinant

$$D = \begin{vmatrix} v_{11} & v_{21} & \cdots & v_{m1} \\ v_{12} & v_{22} & \cdots & v_{m2} \\ \cdot\;\cdot\;\cdot & & & \cdot \\ \cdot\;\cdot\;\cdot & & & \cdot \\ \cdot\;\cdot\;\cdot & & & \cdot \\ v_{1m} & v_{2m} & \cdots & v_{mm} \end{vmatrix}$$

and C_{jj} is the cofactor of the element in the j-th row and j-th column of D. A proof of Eq. 1.41 is given in Chapter 2 (Eq. 2.36).

Expanded Subsystems of Pools

Obviously, if the rates of transfer of material in a system of m primary pools are known, the rates of exchange among the pools in any system consisting of a subset of these pools (e.g., $m-h$) can be calculated. Thus, if the pools are renumbered so that those excluded have indices $m-h+1$ to m (which can be done with no loss of generality), then

$$v_{ji}^{(m-h)} = v_{ji}^{(m)} + \sum_{p=m-h+1}^{m} \Delta_{jp}^{(m-h+1)}\, v_{pi}^{(m)} \qquad \begin{array}{l} i = 1, 2, \ldots, m-h \\ j = 0, 1, \ldots, m-h. \end{array} \tag{1.42}$$

Proof of Eq. 1.42 is given in the next chapter (Eq. 2.44). In particular, if $j = i$ in Eq. 1.42,

$$-v_{ii}^{(m-h)} = -v_{ii}^{(m)} - \sum_{p=m-h+1}^{m} \Delta_{ip}^{(m-h+1)}\, v_{pi}^{(m)}. \tag{1.43}$$

When h is considered to equal m−1, then m−h=1, and the rate $-v_{ii}^{(m-h)} = -v_{ii}^{(1)}$ corresponds, according to previous definitions, to $v_{i0}^{(1)}$ or PR_1 (see Fig. 1.1). In this case, it follows from Eq. 1.43 that

$$PR_1 = -v_{11}^{(m)} - \sum_{p=2}^{m} \Delta_{1p}^{(2)} v_{p1}^{(m)}. \tag{1.44}$$

This expression indicates that the production rate of a pool equals the total rate at which material enters the pool minus that part of the transfers from other pools to pool 1 that corresponds to material returning to pool 1 (as before the rate of recycle w_j is self-canceling and may be ignored). As indicated previously, $\Delta_{1p}^{(2)} v_{p1}^{(m)}$ represents the rate at which material that has already been in pool 1 is transferred from p to 1 without the intermediacy of any other primary pool.

The reader may verify that if m=3 and m−h=2 in Eq. 1.41, then, for j=1 and i=2,

$$v_{12}^{(2)} = v_{12}^{(3)} + \Delta_{13}^{(3)} v_{32}^{(3)}. \tag{1.45}$$

The equivalence between a subsystem including pools 1 and 2 and an expanded subsystem that includes pools 1, 2, and 3 is shown in Fig. 1.8. This equivalence can be verified by comparing the values of rates and Δ fractions corresponding to E_1, E_2, and E_1S, shown in Figs. 1.5 and 1.7 and in Table 1.2.

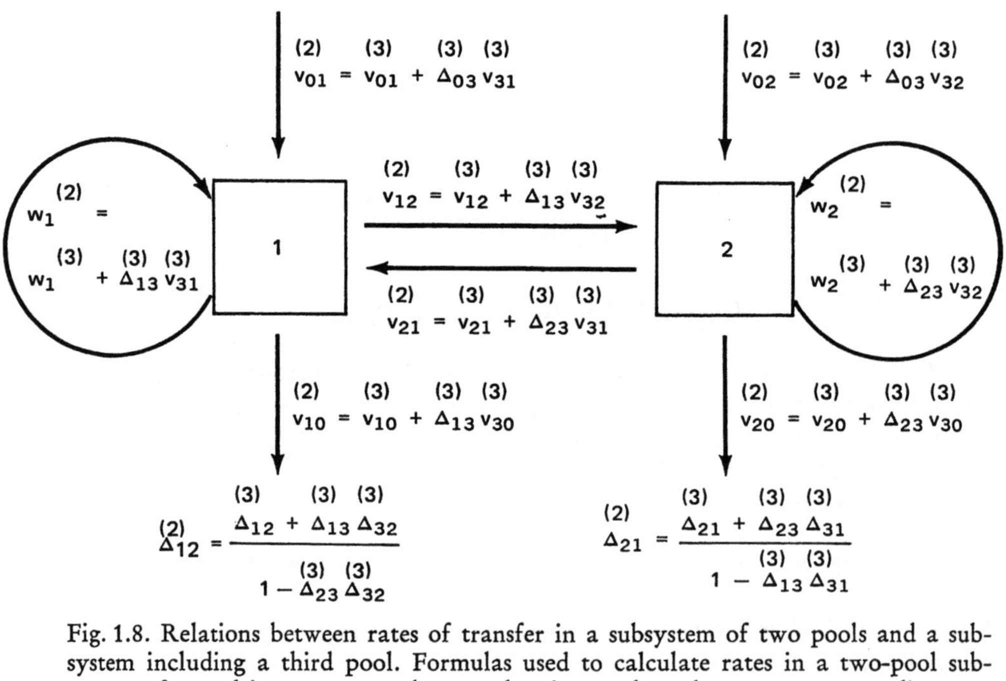

Fig. 1.8. Relations between rates of transfer in a subsystem of two pools and a subsystem including a third pool. Formulas used to calculate rates in a two-pool subsystem of a multicompartmental network using as data the rates corresponding to a subsystem that includes the same two pools and a third pool. These formulas are a particular application of the more general expressions 1.42 to 1.48

If $j=0$ in Eq. 1.42, it follows that

$$v_{0i}^{(m-h)} = v_{0i}^{(m)} + \sum_{p=m-h+1}^{m} \Delta_{0p}^{(m-h+1)} v_{pi}^{(m)} . \tag{1.46}$$

In particular, since $v_{01}^{(1)} = PR_1$ by definition, it results from Eq. 1.46 that if $m-h=1$, then,

$$PR_1 = v_{01}^{(m)} + \sum_{p=2}^{m} \Delta_{0p}^{(2)} v_{p1}^{(m)} . \tag{1.47}$$

This equation indicates that the rate of material entering *de novo* into a pool (PR_1) equals the sum of the rate at which material enters from the outside directly into the pool ($v_{01}^{(m)}$) plus the rates of new material transferring from each of the other pools in the system to pool 1. Because $v_{p1}^{(m)}$ is the total rate of transfer from pool p to 1 without the intermediacy of any other primary pool and $\Delta_{0p}^{(2)}$ is the contribution to pool p of material that entered the system and reached pool p without passing through pool 1, it is clear that $\Delta_{0p}^{(2)} v_{p1}^{(m)}$ is the component of the rate of transfer from p to 1 corresponding to the material that has never been in pool 1.

A summary of formulas relating the production rate of the hormone in a pool to other rates can now be given.

$$PR_1 = v_{01}^{(m)} + \sum_{i=2}^{m} v_{i1}^{\prime(m)} = v_{01}^{(m)} + \sum_{i=2}^{m} \varrho_{i1} v_{0i}^{(m)}$$

$$= v_{01}^{(m)} + \sum_{i=2}^{m} \Delta_{0i}^{(2)} v_{i1}^{(m)} = v_{10}^{(m)} + \sum_{i=2}^{m} \Delta_{1i}^{(2)} v_{i0}^{(m)}$$

where

$$\varrho_{ij} = \frac{v_{ij}^{(2)}}{v_{i0}^{(2)} + v_{ij}^{(2)}} \qquad i, j = 1, 2, \ldots, m, \qquad i \neq j$$

and

$$\Delta_{ij}^{(2)} = \frac{v_{ij}^{(2)}}{v_{0j}^{(2)} + v_{ij}^{(2)}} \qquad \begin{array}{l} i = 0, 1, \ldots, m \\ j = 1, 2, \ldots, m, \quad i \neq j . \end{array}$$

Equations 1.42 and 1.43 are used to derive the relationship between "contribution factors" in a system and a subsystem of primary pools, as follows:

$$\Delta_{ji}^{(m-h)} = \frac{v_{ji}^{(m-h)}}{-v_{ii}^{(m-h)}} = \frac{v_{ji}^{(m)} + \sum_{p=m-h+1}^{m} \Delta_{jp}^{(m-h+1)} v_{pi}^{(m)}}{-v_{ii}^{(m)} - \sum_{p=m-h+1}^{m} \Delta_{ip}^{(m-h+1)} v_{pi}^{(m)}} .$$

If numerator and denominator in this equation are divided by $-v_{ii}^{(m)}$, the following expression is obtained:

$$\Delta_{ji}^{(m-h)} = \frac{\Delta_{ji}^{(m)} + \sum_{p=m-h+1}^{m} \Delta_{jp}^{(m-h+1)} \Delta_{pi}^{(m)}}{1 - \sum_{p=m-h+1}^{m} \Delta_{ip}^{(m-h+1)} \Delta_{pi}^{(m)}} . \tag{1.48}$$

The general expressions derived in this section are of theoretical interest, but their applicability to estimate rates in systems consisting of more than three to four primary pools is questionable. The more tracers that are administered and pools that are sampled, the more detailed is the information obtained about the origin and fate of the compounds under study (each tracer used allows the inclusion of one pool in the system). However, the number of tracers that can be used simultaneously is limited by the number of isotopes available for labeling. Even if consecutive experiments were possible or experiments performed on different occasions could be combined, rates in systems of more than a few primary pools may not be determined readily because of the large number of experimental data involved in their calculations.

Chapter 2

Infusion of Tracers at a Constant Rate

The purpose of this chapter is to introduce formulas that relate the rates described in the preceding chapter to isotopic data obtained during constant infusion of tracers. The derivation of these formulas follows the order of increasing complexity of the systems; accordingly, 1, 2, 3, and m primary pools embedded in a multicompartmental network are considered sequentially. Although the formulas derived for the m pool system are of course applicable when m is 1, 2, or 3, the simpler cases are discussed separately since the symbolism is less cumbersome and, in practice, only systems consisting of very few primary pools can be studied experimentally.

If a tracer is infused at a constant rate into a pool in an open system that is at the steady state, the amount of isotope in each pool reaches a constant level when the same amount of isotope enters and leaves the pool. Although theoretically this isotopic steady state is reached only at infinity, constant values for the amount of isotope in each pool are expected after a finite time of infusion, due to the limit of precision in the measurements.

Isotope measurements are expressed as number of desintegrations per minute (dpm); specific activities (denoted by the symbol a, for "activity") are expressed as dpm/μmole of the compound in the pool under study; infusion rates (I^* for "infusion"), in dpm/hr.

All the formulas in this chapter are obtained from linear algebraic equations describing the isotopic steady state at which the amounts of an isotope entering and leaving the pool are equal.

A. Multicompartmental System in Which Only One Pool is Studied (Fig. 1.1)

If a tracer is introduced at the constant rate of I_1^* dpm/hr into the pool under study (pool 1) and a constant specific activity a_1^1 of the compound in that pool is obtained at the steady state, it follows that

$$I_1^* + w_1 a_1^1 = (v_{10} + w_1) a_1^1$$

or

$$I_1^* = v_{10} a_1^1 .$$

This expression indicates that the rate at which the isotope enters the pool "for the first time" equals the rate at which the isotope leaves the pool irreversibly. Since at the steady state v_{10} equals PR_1, then

$$PR_1 = \frac{I_1^*}{a_1^1}. \qquad (2.1)$$

This result is self-evident since the specific activity of the pool, at the steady state, is determined by the ratio of the rate at which the isotope enters the pool for the first time to the rate at which new material enters the pool.

Defining a metabolic clearance rate (MCR) as the ratio PR_1/c_1 (TAIT, 1963) and noting that the specific activity in the pool equals the isotope concentration divided by the concentration of unlabeled material in the pool ($a_1^1 = c_1^1/c_1$), it follows that

$$MCR_1 = \frac{PR_1}{c_1} = \frac{PR_1\, a_1^1}{c_1^1} = \frac{I_1^*}{c_1^1}. \qquad (2.2)$$

PR's and MCR's, important parameters characterizing blood-borne compounds, are discussed in detail in Chapter 5.

The value of the rate of reentry of the compound into the pool (w_1, Fig. 1.1) is not calculated from isotopic steady-state data. It is possible, however, to estimate w_1 from isotopic data obtained by measuring the concentrations of tracer in the pool after beginning the infusion (transient phase) or following a sudden introduction of the tracer into the pool, as is described in Chapter 3.

B. Multicompartmental System in Which Two Pools are Studied (Fig. 1.2)

If a tracer is infused at a constant rate into pool 1 while another tracer, labeled with a different isotope, is simultaneously infused into pool 2, a set of four constant specific activities (two isotopes and two pools) can be obtained as isotopic data. These four specific activities are designated by the symbols a_1^1, a_2^1, a_1^2, and a_2^2. The superscript indicates the isotope by referring to the pool in which the isotope is introduced; the subscript indicates the pool in which the specific activity is determined. Thus, a_2^1 denotes the specific activity of pool 2 with respect to the isotope infused into pool 1. If, for instance, a tritium-labeled tracer is infused into pool 1, a_2^1 is $a_2^{^3H}$. The rates of infusion of tracers into pools 1 and 2 are denoted by I_1^* and I_2^*, respectively, to indicate the amount of isotope introduced per unit of time (dpm/hr).

Rates

Isotope 1 enters pool 1 from outside the system at the rate I_1^*, from pool 2 at the rate $v_{21}\, a_2^1$, and by recycling processes that do not involve pool 2, at the rate $w_1 a_1^1$. Isotope 1 leaves the pool at the rate $v_{12} a_1^1 + v_{10} a_1^1 + w_1 a_1^1$. Therefore, the isotope steady-state condition implies that

$$I_1^* + v_{21} a_2^1 + w_1 a_1^1 = v_{12} a_1^1 + v_{10} a_1^1 + w_1 a_1^1 \tag{2.3}$$

or

$$(v_{10} + v_{12}) a_1^1 - v_{21} a_2^1 = I_1^* . \tag{2.4}$$

Similarly, the isotopic steady state in pool 1 for an isotope introduced into pool 2 yields the equation

$$(v_{10} + v_{12}) a_1^2 - v_{21} a_2^2 = 0 . \tag{2.5}$$

The system of these two linear simultaneous equations can be solved algebraically to obtain v_{21} and the quantity $v_{10} + v_{12}$ in terms of the specific activities and the infusion rate. Using Cramer's rule,

$$v_{21} = \begin{vmatrix} a_1^1 & -I_1^* \\ a_1^2 & 0 \end{vmatrix} \cdot/\cdot \begin{vmatrix} a_1^1 & a_2^1 \\ a_1^2 & a_2^2 \end{vmatrix} = I_1^* \frac{a_1^2}{a_1^1 a_2^2 - a_2^1 a_1^2} \tag{2.6}$$

and

$$v_{10} + v_{12} = - \begin{vmatrix} -I_1^* & a_2^1 \\ 0 & a_2^2 \end{vmatrix} \cdot/\cdot \begin{vmatrix} a_1^1 & a_2^1 \\ a_1^2 & a_2^2 \end{vmatrix} = I_1^* \frac{a_2^2}{a_1^1 a_2^2 - a_2^1 a_1^2} . \tag{2.7}$$

Similarly, the isotope balance at the steady state for pool 2 is described by the following equations:

$$(v_{20} + v_{21}) a_2^1 - v_{12} a_1^1 = 0 \tag{2.8}$$
$$(v_{20} + v_{21}) a_2^2 - v_{12} a_1^2 = I_2^* . \tag{2.9}$$

Therefore,

$$v_{12} = I_2^* \frac{a_2^1}{a_1^1 a_2^2 - a_2^1 a_1^2} \tag{2.10}$$

and

$$v_{20} + v_{21} = I_2^* \frac{a_1^1}{a_1^1 a_2^2 - a_2^1 a_1^2} . \tag{2.11}$$

The rates v_{10} and v_{20} are calculated from the same data since, obviously,

$$v_{10} = (v_{10} + v_{12}) - v_{12} \qquad \text{and} \qquad v_{20} = (v_{20} + v_{21}) - v_{21} .$$

The rates v_{01} and v_{02} are calculated from the steady-state equations 1.13 and 1.14.

These expressions are included in Table 2.1 where several formulas used in the calculation of rates from isotopic steady-state data are listed.

Equations 2.6 to 2.11 are valid only if the determinant

$$D = \begin{vmatrix} a_1^1 & a_2^1 \\ a_1^2 & a_2^2 \end{vmatrix} = a_1^1 a_2^2 - a_2^1 a_1^2$$

does not equal zero. That the determinant is not zero in an open system of pools with nonzero inputs is shown by multiplying D by another determinant, as follows (Gurpide *et al.*, 1963 b):

$$\begin{vmatrix} a_1^1 & a_2^1 \\ a_1^2 & a_2^2 \end{vmatrix} \cdot \begin{vmatrix} v_{11} & v_{12} \\ v_{21} & v_{22} \end{vmatrix} = \begin{vmatrix} (v_{11} a_1^1 + v_{21} a_2^1) & (v_{12} a_1^1 + v_{22} a_2^1) \\ (v_{11} a_1^2 + v_{21} a_2^2) & (v_{12} a_1^2 + v_{22} a_2^2) \end{vmatrix}$$

Table 2.1. Formulas used to calculate rates of formation, transfer, and removal in a two pool subsystem (Fig. 1.2) from isotopic steady-state data [a]

$$PR_1 = \frac{I_1^*}{a_1^1} \qquad PR_2 = \frac{I_2^*}{a_2^2}$$

$$\varrho_{12} = \frac{I_2^* a_2^1}{I_1^* a_2^2} \qquad \varrho_{21} = \frac{I_1^* a_1^2}{I_2^* a_1^1}$$

$$\Delta_{12} = \frac{a_2^1}{a_1^1} \qquad \Delta_{21} = \frac{a_1^2}{a_2^2}$$

$$\text{Denominator} = D = a_1^1 a_2^2 - a_2^1 a_1^2$$

$$v_{01} = I_1^* \frac{a_2^2 - a_1^2}{D} \qquad v_{02} = I_2^* \frac{a_1^1 - a_2^1}{D}$$

$$v_{12} = I_2^* \frac{a_2^1}{D} \qquad v_{21} = I_1^* \frac{a_1^2}{D}$$

$$v_{10} = \frac{I_1^* a_2^2 - I_2^* a_2^1}{D} \qquad v_{20} = \frac{I_2^* a_1^1 - I_1^* a_1^2}{D}$$

[a] I_1^* and I_2^* denote the rates of infusion (dpm/hr) of tracers into pools 1 and 2, respectively. The specific activities at the steady state are denoted by the symbols a_1^1, a_2^1, a_1^2, and a_2^2. The subscript indicates the pool. The superscript indicates the isotope by referring to the pool through which it was introduced into the system.

where the symbols $v_{11} = -(v_{10} + v_{12})$ and $v_{22} = -(v_{20} + v_{21})$ (Eq. 1.33) have been introduced in order to simplify the expressions. Each element of the product determinant can be replaced by $-I_1^*$, $-I_2^*$ or 0 since, using the notation v_{11} and v_{22}, Eqs. 2.4, 2.5, 2.8, and 2.9 are equivalent to

$$\begin{aligned} v_{11} a_1^1 + v_{21} a_2^1 &= -I_1^* \\ v_{11} a_1^2 + v_{21} a_2^2 &= 0 \\ v_{12} a_1^1 + v_{22} a_2^1 &= 0 \\ v_{12} a_1^2 + v_{22} a_2^2 &= -I_2^* . \end{aligned}$$

Thus,

$$\begin{vmatrix} a_1^1 & a_2^1 \\ a_1^2 & a_2^2 \end{vmatrix} \begin{vmatrix} v_{11} & v_{12} \\ v_{21} & v_{22} \end{vmatrix} = \begin{vmatrix} -I_1^* & 0 \\ 0 & -I_2^* \end{vmatrix} = I_1^* I_2^* \neq 0 . \tag{2.12}$$

Because the product of the two determinants does not equal zero, neither of these determinants can be zero, i.e., $D \neq 0$.

It should be noted, however, that if the experimental value of D is very low in comparison to the value of the numerator in the expressions used to calculate the rates (e.g., Eq. 2.6), large errors in the calculated values can be expected and the experimental design may prove inadequate. Such a situation may arise if the rates of exchange among the pools are extremely fast relative to the irreversible loss rates. In that case, the values of the specific activity with respect to any given isotope will tend to be identical; thus the columns of the specific activity matrix will be very similar. By the rules of algebra, subtraction of one column from another in the determinant D does not change the value of the determinant. As this will produce a column consisting of terms approaching 0, it is clear that under such conditions the value of the determinant will be near zero.

It is of practical importance from an experimental viewpoint to note that the determination of a rate of transfer of material from one pool to another requires measurement of specific activities in one of the pools but only isotope ratio in the other. This property can be verified by dividing numerator and denominator in Eq. 2.6 by the specific activity appearing in the numerator (a_1^2). Thus,

$$v_{21} = \frac{I_1^*}{R_1 a_2^2 - a_2^1} \tag{2.13}$$

where R_1 denotes the isotope ratio (isotope 1/isotope 2) in the compound in pool 1. No specific activities of the compound in this pool appear in the formula.

Similarly,

$$v_{12} = \frac{I_2^*}{a_1^1 (1/R_2) - a_1^2}$$

where R_2 is the isotope ratio (isotope 1/isotope 2) in the compound in pool 2.

It is experimentally simpler to measure isotope ratios than specific activities, particularly when the concentrations of the hormone are very low. Addition of an unlabeled compound (carrier) to the sample does not affect isotope ratios and facilitates the isolation procedures.

Transfer Factors (ϱ)

The transfer factor was defined in the preceding chapter (Eq. 1.6) as $\varrho_{12} = v_{12}/(v_{10} + v_{12})$. It is now a simple procedure to use Eqs. 2.7 and 2.10 to derive the following formula for the estimation of ϱ_{12} from isotopic data:

$$\varrho_{12} = \frac{I_2^*}{I_1^*} \frac{a_2^1}{a_2^2} = \frac{R_2}{R \text{ infused}} \, . \tag{2.14}$$

Similarly, by dividing Eq. 2.6 by Eq. 2.11, it follows that

$$\varrho_{21} = \frac{I_1^*}{I_2^*} \frac{a_1^2}{a_1^1} = \frac{R \text{ infused}}{R_1} \, . \tag{2.15}$$

These equations indicate that the ϱ factors are actually given in terms of isotope ratios; measurement of specific activities is unnecessary. In fact, the simplicity of these measurements motivated the search for the correct rate equation to be assigned to such experimental values. The use of "transfer factors" in endocrinology originally appeared in studies of the irreversible metabolism of epinephrine (Kopin, 1960). The application of the ϱ's to reversible reactions was introduced during the course of studies of the interconversion of steroids (Gurpide *et al.*, 1963 a).

It should be recalled that in Chapter 1 it was stated that all the v's in a two-pool subsystem could be calculated if PR_1, PR_2, ϱ_{12}, and ϱ_{21} were known. The corresponding relationships were listed in Table 1.1. The reader can now verify these relations. The pertinent general proofs are provided in another section of this chapter (see Eq. 2.30).

It is important to establish a connection between the rate of infusion of a tracer into a pool and the resulting steady-state specific activity of compounds in other pools. Since $PR_1 = I_1^*/a_1^1$, it follows from Eq. 2.15 that

$$a_1^2 = \frac{\varrho_{21}\, I_2^*}{PR_1} \tag{2.16}$$

or

$$PR_1 = \frac{\varrho_{21}\, I_2^*}{a_1^2}\,. \tag{2.17}$$

These results indicate that as far as pool 1 is concerned, an infusion of I_2^* dpm/hr into pool 2 is equivalent to an infusion of $\varrho_{21}\, I_2^*$ dpm/hr into pool 1.

Contribution Factors (Δ)

The fraction Δ_{12} corresponding to a two-primary-pool system has been defined in the preceding chapter (Eq. 1.15) as $\Delta_{12} = v_{12}/(v_{02}+v_{12}) = v_{12}/(v_{20}+{}_{21})$. Since the steady state in pool 2 can be expressed by Eq. 2.8, viz.

$$v_{12}\, a_1^1 = (v_{20}+v_{21})\, a_2^1$$

it follows by dividing both sides of the equation by $(v_{20}+v_{21})$ that $\Delta_{12}\, a_1^1 = a_2^1$, i.e.,

$$\Delta_{12} = \frac{a_2^1}{a_1^1}\,. \tag{2.18}$$

Rates of Transfer Which Exclude Recycle

In Chapter 1, the rate v_{12} was dissected into two component rates v'_{12} and v''_{12}, which correspond, respectively, to material that has never been in pool 2 and material that is returning to pool 2. By definition, the material included in the rate v''_{12} has, at the isotopic steady state, the same specific activity with respect to the isotope infused into pool 2 as the material in pool 2, because all of it originates in that pool. Therefore, the amount of isotope 2 transferred from pool 1 to pool 2, which equals $v_{12}\, a_1^2$, can also be expressed as $v''_{12}\, a_2^2$. Then,

$$v_{12}\, a_1^2 = v''_{12}\, a_2^2$$

or

$$v''_{12} = \frac{a_1^2}{a_2^2}\, v_{12} = \Delta_{21}\, v_{12}\,. \tag{2.19}$$

Since $v'_{12} = v''_{12}$, it follows that v'_{12}, v''_{12}, (and also v'_{21} and v''_{21}) can also be calculated from specific activities in pools 1 and 2.

The reader may now verify Eq. 1.5 in Chapter 1 by expressing ϱ_{12}, v''_{12}, and v_{21} in terms of specific activities. Note the usefulness of these substitution procedures in the search for relationships between kinetic parameters.

C. Multicompartmental System in Which Three Pools are Studied (Fig. 1.6)

The formulas for the calculation of the rates of transfer and removal of material in a three-primary-pool system are derived in a manner identical to that described for the two-primary-pool system.

The experimental data necessary for the estimation of these rates include the determination of specific activities of three pools during infusion of three tracers. In practice, two sets of separate experiments using different pairs of tracers can be conducted. For instance, $I_1^{^3H}$ and $I_2^{^{14}C}$ may be infused in one experiment, while $I_2^{^{14}C}$ and $I_3^{^3H}$ may be infused in the other. Pools 1, 2, and 3 must be sampled in each experiment.

The notation $v_{11} = -(v_{10} + v_{12} + v_{13})$, $v_{22} = -(v_{20} + v_{21} + v_{23})$ and $v_{33} = -(v_{30} + v_{31} + v_{32})$ is used in order to simplify the equations that follow.

Rates

The conditions of steady state in pool 1 with respect to isotopes introduced into pools 1, 2, and 3 are expressed by the following set of equations:

$$\begin{aligned} v_{11}\,a_1^1 + v_{21}\,a_2^1 + v_{31}\,a_3^1 &= -I_1^* \\ v_{11}\,a_1^2 + v_{21}\,a_2^2 + v_{31}\,a_3^2 &= 0 \\ v_{11}\,a_1^3 + v_{21}\,a_2^3 + v_{31}\,a_3^3 &= 0\,. \end{aligned} \qquad (2.20)$$

The solution of this system of three simultaneous equations with three unknowns yields formulas from which the rates v_{11}, v_{21}, and v_{31} can be estimated from the specific activities at the steady state.

Similar sets of equations pertaining to pools 2 and 3 yield expressions from which the rates corresponding to these pools can be calculated.

The rates of entry of material into each pool from outside the system and the rates of irreversible removal of material from the system are then obtained using steady-state equations. For example,

$$v_{01} = (-v_{11}) - v_{21} - v_{31} \qquad \text{and} \qquad v_{10} = (-v_{11}) - v_{12} - v_{13}\,.$$

The formulas giving rates in terms of specific activities are not explicitly presented (see Table 1.3, which shows expressions that can be used to estimate these rates on the basis of PR and ϱ values).

Contribution Factors

The factor Δ_{13} corresponding to a three-primary-pool system, is defined in Chapter 1 (Eq. 1.28) as the ratio $v_{13}/(-v_{33})$. An expression for Δ_{13} in terms of specific activities can be obtained by substituting the values of the rates v_{13} and $-v_{33}$ calculated by solving a system of equations similar to Eq. 2.20 obtained by considering the isotope balance in pool 3, at the steady state. An identical formula for Δ_{13} can be more directly derived on the basis of the following two equations, which describe the steady-state situation in pool 3 with respect to the isotopes introduced into pools 1 and 2:

$$v_{13}\, a_1^1 + v_{23}\, a_2^1 = -v_{33}\, a_3^1$$
$$v_{13}\, a_1^2 + v_{23}\, a_2^2 = -v_{33}\, a_3^2 .$$

If both equations are divided by $-v_{33}$, then

$$\Delta_{13}\, a_1^1 + \Delta_{23}\, a_2^1 = a_3^1$$
$$\Delta_{13}\, a_1^2 + \Delta_{23}\, a_2^2 = a_3^2 .$$

From this system of simultaneous equations, it can be calculated that

$$\Delta_{13} = \begin{vmatrix} a_3^1 & a_2^1 \\ a_3^2 & a_2^2 \end{vmatrix} \cdot/\cdot \begin{vmatrix} a_1^1 & a_2^1 \\ a_1^2 & a_2^2 \end{vmatrix} = \frac{a_3^1 a_2^2 - a_3^2 a_2^1}{a_1^1 a_2^2 - a_2^1 a_1^2} . \qquad (2.21)$$

Note that when the numerator and denominator of this equation is divided by a_2^2, it follows that

$$\Delta_{13} = \frac{a_3^1 - a_3^2\,(a_2^1/a_2^2)}{a_1^1 - a_1^2\,(a_2^1/a_2^2)} = \frac{a_3^1 - R_2\, a_3^2}{a_1^1 - R_2\, a_1^2} \qquad (2.22)$$

where, as before, R_2 is the ratio of isotopes 1 and 2 in pool 2, at the steady state.

Furthermore, if 1 and 2 are the only sources of material in pool 3 ($\Delta_{13} + \Delta_{23} = 1$) then

$$a_3^1 = \Delta_{13}\, a_1^1 + (1 - \Delta_{13})\, a_2^1$$
$$a_3^2 = \Delta_{13}\, a_1^2 + (1 - \Delta_{13})\, a_2^2$$

and

$$R_3 = \frac{\Delta_{13}\,(a_1^1 - a_2^1) + a_2^1}{\Delta_{13}\,(a_1^2 - a_2^2) + a_2^2}$$

and, making Δ_{13} explicit,

$$\Delta_{13} = \frac{1}{1 + \dfrac{R_1 - R_3}{R_3 - R_2}\,\dfrac{a_1^2}{a_2^2}} . \qquad (2.23)$$

"Non Recycling" Transfer Rates

The same reasoning followed to derive Eq. 2.19 can be applied to calculate rates such as $\dot{v}_{13}^{\prime(3)}$ and $v_{13}^{\prime\prime(3)}$ (Chapter 1) in a three-primary-pool system. Thus, using the superscript in parentheses to indicate the number of primary pools in the system,

$$v_{13}^{(3)}\, a_1^3 = v_{13}^{\prime\prime(3)}\, a_3^3$$

and

$$v_{13}^{\prime\prime(3)} = \frac{a_1^3}{a_3^3}\, v_{13}^{(3)} = \Delta_{13}^{(2)}\, v_{13}^{(3)} . \qquad (2.24)$$

Therefore, $v'^{(3)}_{13}$, which describes the rate at which material that has never been in pool 3 is transferred from pools 1 to 3 bypassing pool 2, is given by the following expression:

$$v'^{(3)}_{13} = v^{(3)}_{13} - v''^{(3)}_{13} = (1 - \Delta^{(2)}_{13})\, v^{(3)}_{13}. \tag{2.25}$$

Transfer Factors in a Subsystem of Three Primary Pools Connected in Series (Fig. 2.1)

It follows from the definition of ϱ that in any three-pool subsystem

$$\frac{\varrho_{12} \times \varrho_{23}}{\varrho_{13}} = \frac{\dfrac{I^*_2}{I^*_1}\dfrac{a^1_2}{a^2_2} \times \dfrac{I^*_3}{I^*_2}\dfrac{a^2_3}{a^3_3}}{\dfrac{I^*_3}{I^*_1}\dfrac{a^1_3}{a^3_3}} = \frac{a^1_2\, a^2_3}{a^2_2\, a^1_3}. \tag{2.26}$$

In a three-pool subsystem in which the three primary pools are connected in series (Fig. 2.1) all exchange between pools 1 and 3 occur through pool 2. If

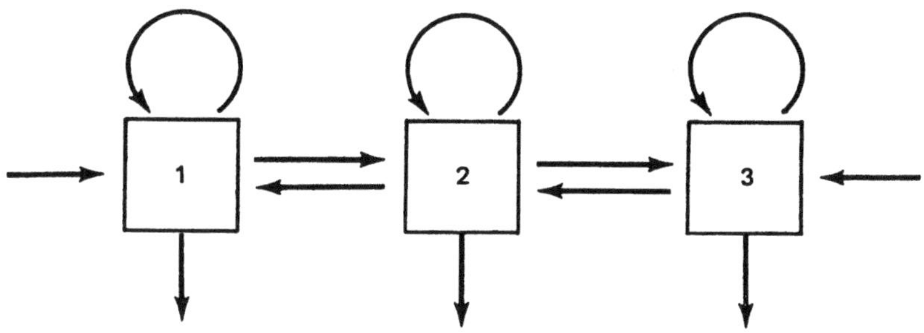

Fig. 2.1. Transfer factors in a subsystem of three pools connected in series. As indicated by Eq. 2.27, $\varrho_{13} = \varrho_{12}\, \varrho_{23}$ and $\varrho_{31} = \varrho_{32}\, \varrho_{21}$ if $v^{(3)}_{13} = v^{(3)}_{31} = 0$

isotopically labeled tracers are infused into pools 1 and 2, the isotope ratios in pools 2 and 3 are identical at the steady state, since all labeled material reaching pool 3 comes from a common source, viz. pool 2. Therefore,

$$\frac{a^1_2}{a^2_2} = \frac{a^1_3}{a^2_3} \quad \text{and} \quad \frac{a^1_2\, a^2_3}{a^2_2\, a^1_3} = 1$$

and, from Eq. 2.26,

$$\varrho_{13} = \varrho_{12}\, \varrho_{23} \qquad \text{if } v^{(3)}_{13} = 0\,. \tag{2.27}$$

It should be emphasized that this relation is not applicable if there is a pathway of transfer from pool 1 to 3 not involving pool 2.

D. General m-Primary-Pool Subsystem

In a subsystem of m-primary pools, m may vary from 1 to n, since n is used to denote the total number of pools present in the system, a number that is conveniently left unspecified because it usually cannot be ascertained.

Rates

The expression describing the condition of steady state in a pool (pool j) with respect to an isotope introduced at a constant rate (into pool k) is the following:

$$\sum_{i=1}^{m} v_{ij}\, a_i^k = -\delta_{kj}\, I_k^* \qquad k, j = 1, 2, \ldots, m \tag{2.28}$$

where δ_{kj} is a coefficient that equals 0 if $k \neq j$ and equals unity if $k = j$.

This equation states that the amount of isotope leaving the pool ($-v_{jj}\, a_j^k$) equals the amount of isotope entering the pool ($\sum_{i=1, i \neq j}^{m} v_{ij}\, a_i^k$) plus, if $k = j$, the rate of infusion of the tracer (I_j^*). The rate v_{0j} does not appear in the equation because, by definition, no isotope is associated with the rate of entry of material into pool j from outside the system. In addition, the rate w_j does not appear in Eq. 2.28 because, also by definition, as much isotope (viz. $w_j\, a_j^k$) leaves the pool and returns to it by the processes denoted by this rate, at the steady state.

A system of m simultaneous linear equations is obtained by assigning to k in Eq. 2.28, the values 1, 2, ..., m, i.e.,

$$\begin{aligned} v_{1j}\, a_1^1 + v_{2j}\, a_2^1 + \cdots + v_{mj}\, a_m^1 &= 0 \\ \vdots \qquad \vdots \qquad\quad &\vdots \\ v_{1j}\, a_1^j + v_{2j}\, a_2^j + \cdots + v_{mj}\, a_m^j &= -I_j^* \\ \vdots \qquad \vdots \qquad\quad & \\ v_{1j}\, a_1^m + v_{2j}\, a_2^m + \cdots + v_{mj}\, a_m^m &= 0\,. \end{aligned} \tag{2.29}$$

The values of the m unknowns, v_{ij}, $i = 1, 2, \ldots, m$, can then be calculated from the specific activities in each of the m pools after the administration of m tracers, each in one pool (simultaneously or in different experiments). The same set of data is used to calculate the rates corresponding to each of j pools, where $j = 1, 2, \ldots, m$.

The rates v_{0j}, $j = 1, 2, \ldots, m$, are then obtained using Eq. 1.35, which states that $v_{0j}^{(m)} = -\sum_{i=1}^{m} v_{ji}^{(m)}$.

It is evident that the value of any rate v_{ij} can only be calculated if the determinant of the coefficients, in the set of equations 2.29, is not equal to zero, i.e.,

$$A = \begin{vmatrix} a_1^1 & a_2^1 & \ldots & a_m^1 \\ a_1^2 & a_2^2 & \ldots & a_m^2 \\ \vdots & \vdots & & \vdots \\ a_1^m & a_2^m & \ldots & a_m^m \end{vmatrix} \neq 0\,.$$

Using rules of multiplication of determinants (WADE, 1951) and Eq. 2.29, it can be verified that, similarly to Eq. 2.12,

$$A \times \begin{vmatrix} v_{11} & v_{12} & \cdots & v_{1m} \\ v_{21} & v_{22} & \cdots & v_{2m} \\ \vdots & \vdots & & \vdots \\ v_{m1} & v_{m2} & \cdots & v_{mm} \end{vmatrix} = \begin{vmatrix} -I_1^* & 0 & \cdots & 0 \\ 0 & -I_2^* & \cdots & 0 \\ \vdots & \vdots & & \\ 0 & 0 & \cdots & -I_m^* \end{vmatrix} = (-1)^m I_1^* I_2^* \ldots I_m^* \neq 0 .$$

This result assures the consistency of Eqs. 2.29.

Rates in Terms of PR's and ϱ's

If each specific activity in Eq. 2.29 is replaced by expressions similar to Eq. 2.16 ($a_i^k = \varrho_{ki} I_k^*/PR_i$), it follows that

$$\begin{aligned} I_1^* \left(v_{1j} \frac{1}{PR_1} + v_{2j} \frac{\varrho_{12}}{PR_2} + \cdots + v_{mj} \frac{\varrho_{1m}}{PR_m}\right) &= 0 \\ \vdots \qquad \qquad \vdots \qquad \qquad & \vdots \\ I_j^* \left(v_{1j} \frac{\varrho_{j1}}{PR_1} + v_{2j} \frac{\varrho_{j2}}{PR_2} + \cdots + v_{mj} \frac{\varrho_{jm}}{PR_m}\right) &= -I_j^* \\ \vdots \qquad \qquad \vdots \qquad \qquad & \vdots \\ I_m^* \left(v_{1j} \frac{\varrho_{m1}}{PR_1} + v_{2j} \frac{\varrho_{m2}}{PR_2} + \cdots + v_{mj} \frac{1}{PR_m}\right) &= 0 \end{aligned}$$

or

$$\begin{aligned} \frac{1}{PR_1} v_{1j} + \frac{\varrho_{12}}{PR_2} v_{2j} + \cdots + \frac{\varrho_{1m}}{PR_m} v_{mj} &= 0 \\ \vdots \qquad \qquad \vdots \qquad \qquad & \vdots \\ \frac{\varrho_{j1}}{PR_1} v_{1j} + \frac{\varrho_{j2}}{PR_2} v_{2j} + \cdots + \frac{\varrho_{jm}}{PR_m} v_{mj} &= -1 \\ \vdots \qquad \qquad \vdots \qquad \qquad & \vdots \\ \frac{\varrho_{m1}}{PR_1} v_{1j} + \frac{\varrho_{m2}}{PR_2} v_{2j} + \cdots + \frac{1}{PR_m} v_{mj} &= 0 . \end{aligned} \tag{2.30}$$

This system of equations in the unknowns $v_{1j}, v_{2j}, \ldots, v_{mj}$ can be conveniently solved by the method of determinants using Cramer's rule. A number m of different systems of equations are obtained by making $j = 1, 2, \ldots, m$. All rates, and therefore all contribution factors, can be calculated from data on production rates and transfer factors by solving these systems of equations. Such calculations are illustrated in the example shown in Fig. 1.7 and Table 1.2.

Contribution Factors

The factor $\Delta_{ij}^{(m)}$, defined in Chapter 1 as the ratio $v_{ij}^{(m)}/(-v_{jj}^{(m)})$, $i \neq j$, $j \neq 0$ can be estimated from the specific activities of the m pools following the infusion of tracers in each of these pools with the exception of pool j. The formulas relating $\Delta_{ij}^{(m)}$ with the isotopic data can be derived from the following equation, included in Eq. 2.28.

$$\sum_{i=1}^{m} v_{ij}\, a_i^k = 0 \qquad k = 1, 2, \ldots, m, \qquad k \neq j\,. \tag{2.31}$$

If the term v_{jj} is taken out of the summation sign, Eq. 2.31 can be written as

$$\sum_{\substack{i=1\\ i\neq j}}^{m} v_{ij}\, a_i^k = -v_{jj}\, a_j^k \qquad k = 1, 2, \ldots, m, \qquad k \neq j\,.$$

Dividing this equation by $-v_{jj}$ it follows from the definition of the Δ's that

$$a_j^k = \sum_{\substack{i=1\\ i\neq j}}^{m} \Delta_{ij}^{(m)}\, a_i^k \qquad k = 1, 2, \ldots, m, \qquad k \neq j\,. \tag{2.32}$$

Therefore, a system of $m-1$ linear simultaneous equations (one for each value of k) is available for the calculation of $m-1$ unknowns (the Δ_{ij}, $i = 1, 2, \ldots, m$, $i \neq j$).

The factor $\Delta_{0j}^{(m)}$ defined in Chapter 1 as the contribution to pool j of material entering the pool from outside the system, i.e., $v_{0j}^{(m)}/(-v_{jj})^{(m)}$, is obtained from Eq. 1.38. Thus,

$$\Delta_{0j}^{(m)} = 1 - \sum_{\substack{i=1\\ i\neq j}}^{m} \Delta_{ij}^{(m)}$$

Proof of the Relationship $PR_1 = v_{01} + \varrho_{21}\, v_{02} + \cdots \varrho_{m1}\, v_{0m}$

According to Eq. 2.28, the following system of equations results when $k = 1$:

$$\begin{aligned}
v_{11}\, a_1^1 + v_{21}\, a_2^1 + \cdots + v_{m1}\, a_m^1 &= -I_1^* \\
v_{12}\, a_1^1 + v_{22}\, a_2^1 + \cdots + v_{m2} a_m^1 &= 0 \\
\vdots \qquad \vdots \qquad\qquad \vdots \quad & \\
v_{1m}\, a_1^1 + v_{2m}\, a_2^1 + \cdots + v_{m2}\, a_m^1 &= 0\,.
\end{aligned} \tag{2.33}$$

If D denotes the determinant of the coefficient in the system, i.e.,

$$D = \begin{vmatrix} v_{11} & v_{21} & \cdots & v_{m1} \\ v_{12} & v_{22} & \cdots & v_{m2} \\ \vdots & \vdots & & \vdots \\ v_{1m} & v_{2m} & \cdots & v_{mm} \end{vmatrix}$$

and C_{ij} is the cofactor of the element of D in the i-th row and j-th column, it follows that

$$a_1^1 = -I_1^* \frac{C_{11}}{D}\,. \tag{2.34}$$

Systems of equations similar to 2.33 can be written for any value of k in Eq. 2.28. These systems of equations differ in the specific activities (a's) but not in the coefficients. Therefore,

$$a_1^k = -I_k^* \frac{C_{11}}{D}. \tag{2.35}$$

According to Eq. 2.34,

$$PR_1 = \frac{I_1^*}{a_1^1} = \frac{-D}{C_{11}}. \tag{2.36}$$

Adding to each element in the first column of D the sum of the corresponding elements of all the other columns, it follows from Eq. 1.33 that

$$PR_1 = \frac{I_1^*}{a_1^1} = \frac{-D}{C_{11}} = \frac{1}{C_{11}} \begin{vmatrix} v_{01} & v_{21} & \cdots & v_{m1} \\ v_{02} & v_{22} & \cdots & v_{m2} \\ \vdots & \vdots & & \vdots \\ v_{0m} & v_{2m} & \cdots & v_{mm} \end{vmatrix}$$

or

$$PR_1 = v_{01} \frac{C_{11}}{C_{11}} + v_{02} \frac{C_{21}}{C_{11}} + \cdots + v_{0m} \frac{C_{m1}}{C_{11}}.$$

It follows from Eqs. 2.34 and 2.35 and from the definition of the transfer factor ϱ_{kl} that

$$\varrho_{k1} = \frac{I_1^*}{I_k^*} \frac{a_1^k}{a_1^1} = \frac{C_{k1}}{C_{11}}. \tag{2.37}$$

Therefore,

$$PR_1 = v_{01} + \varrho_{21} v_{02} + \cdots + \varrho_{m1} v_{0m}. \tag{2.38}$$

Relationships Between Rates in a System of m-Primary Pools and a Subsystem of Some of these Pools

Chapter 1 presented relationships between rates in an m-primary-pool system and a subsystem consisting of m − h of the original m pools. Formulas were given to calculate $v_{ij}^{(m-h)}$, $v_{0j}^{(m-h)}$, and $\Delta_{ij}^{(m-h)}$ from the values of the corresponding parameters in the m-pool system. Each of those formulas was based on Eq. 1.41. The proof of that equation is now provided.

According to Eq. 2.28

$$-\delta_{ki} I_k^* = \sum_{j=1}^{m} v_{ji}^{(m)} a_j^k.$$

Since, by definition, $a_j^k = 0$ when $j = 0$, Eq. 2.28 can be written as

$$\delta_{ki} I_k^* = \sum_{j=0}^{m} v_{ji}^{(m)} a_j^k \qquad k, i = 1, 2, \ldots, m \tag{2.39}$$

where $\delta_{ki} = 1$ if $k = i$ and $\delta_{ki} = 0$ if $k \neq i$.

Equation 2.39 can be resolved in the following summation:

$$\delta_{ki}\, I_k^* = \sum_{j=0}^{m-h} v_{ji}^{(m)}\, a_j^k + \sum_{p=m-h+1}^{m} v_{pi}^{(m)}\, a_p^k . \tag{2.40}$$

Let us now consider a system made of the $m-h$ primary pools of the original system plus *one* other primary pool (pool p). Then, if $k \leqq m-h$, i.e., if the isotope is introduced into one of the pools of the original system,

$$-v_{pp}^{(m-h+1)}\, a_p^k = \sum_{j=0}^{m-h} v_{jp}^{(m-h+1)}\, a_j^k \qquad p > m-h$$

or

$$a_p^k = \sum_{j=0}^{m-h} \Delta_{jp}^{(m-h+1)}\, a_j^k . \tag{2.41}$$

where $\Delta_{jp}^{(m-h+1)} = v_{jp}^{(m-h+1)} / -v_{pp}^{(m-h+1)}$ in the system of primary pools $1, 2 \ldots m-h, p$ defined above.

Substitution of the a_p^k from Eq. 2.41 in Eq. 2.40 yields

$$-\delta_{ki}\, I_k^* = \sum_{j=0}^{m-h} v_{ji}^{(m)}\, a_j^k + \sum_{p=m-h+1}^{m} v_{pi}^{(m)} \sum_{j=0}^{m-h} \Delta_{jp}^{(m-h+1)}\, a_j^k$$

or

$$-\delta_{ki}\, I_k^* = \sum_{j=0}^{m-h} v_{ji}^{(m)}\, a_j^k + \sum_{j=0}^{m-h} \sum_{p=m-h+1}^{m} \Delta_{jp}^{(m-h+1)}\, v_{pi}^{(m)}\, a_j^k$$

or

$$-\delta_{ki}\, I_k^* = \sum_{j=0}^{m-h} \left[v_{ji}^{(m)} + \sum_{p=m-h+1}^{m} \Delta_{jp}^{(m-h+1)}\, v_{pi}^{(m)} \right] a_j^k, \quad 1 \leqq k \leqq h \tag{2.42}$$

Equation 2.42 represents a system of $m-h$ linear simultaneous equations each corresponding to a value of k from 1 to $m-h$, i.e.

$$\begin{aligned}
&\left[v_{1i}^{(m)} + \sum_{p=m-h+1}^{m} \Delta_{1p}^{(m-h+1)} v_{pi}^{(m)} \right] a_1^1 + \cdots + \left[v_{m-h,i}^{(m)} + \sum_{p=m-h+1}^{m} \Delta_{m-h,p}^{(m-h+1)} v_{pi}^{(m)} \right] a_{m-h}^1 = 0 \\
&\qquad \vdots \\
&\left[v_{1i}^{(m)} + \sum_{p=m-h+1}^{m} \Delta_{1p}^{(m-h+1)} v_{pi}^{(m)} \right] a_1^i + \cdots + \left[v_{m-h,i}^{(m)} + \sum_{p=m-h+1}^{m} \Delta_{m-h,p}^{(m-h+1)} v_{pi}^{(m)} \right] a_{m-h}^i = -I_i^* \\
&\qquad \vdots \\
&\left[v_{1i}^{(m)} + \sum_{p=m-h+1}^{m} \Delta_{1p}^{(m-h+1)} v_{pi}^{(m)} \right] a_1^{m-h} + \cdots + \left[v_{m-h,i}^{(m)} + \sum_{p=m-h+1}^{m} \Delta_{m-h,p}^{(m-h+1)} v_{pi}^{(m)} \right] a_{m-h}^{m-h} = 0
\end{aligned}$$

If one applies Eq. 2.28 to the original system of $m-h$ primary pools, one obtains

$$-\delta_{ki}\, I_k = \sum_{j=0}^{m-h} v_{ji}^{(m-h)}\, a_j^k, \quad k, i, = 1, 2 \ldots m-h \tag{2.43}$$

which represent a system of $m-h$ simultaneous linear equations, i.e.

$$\begin{aligned}
&v_{1i}^{(m-h)}\, a_1^1 + \cdots + v_{m-h,i}^{(m-h)}\, a_{m-h}^1 = 0 \\
&\quad \vdots \\
&v_{1i}^{(m-h)}\, a_1^i + \cdots + v_{m-h,i}^{(m-h)}\, a_{m-h}^i = -I_i^* \\
&\quad \vdots \\
&v_{1i}^{(m-h)}\, a_1^{m-h} + \cdots + v_{m-h,i}^{(m-h)}\, a_{m-h}^{m-h} = 0
\end{aligned}$$

It can be seen that both sets of Eqs. 2.42 and 2.43, in which the coefficients of the specific activities are considered to be unknowns, give the same solutions for their corresponding unknowns. For instance, the unknown $v_{1i}^{(m)} + \sum_{p=m-h+1}^{m} \Delta_{1p}^{(m-h+1)} v_{pi}$ in Eq. 2.42 calculated by Cramer's rule is expressed by the following formula:

$$v_{1i}^{(m)} + \sum_{p=m-h+1}^{m} \Delta_{1p}^{(m-h+1)} v_{pi} = \begin{vmatrix} 0 & a_2^1 & \dots & a_{m-h}^1 \\ \vdots & \vdots & & \vdots \\ -I_i^* & a_2^i & \dots & a_{m-h}^i \\ \vdots & \vdots & & \vdots \\ 0 & a_2^{m-h} & \dots & a_{m-h}^{m-h} \end{vmatrix} \Big/ \begin{vmatrix} a_1^1 & a_2^1 & \dots & a_{m-h}^1 \\ \vdots & & & \\ a_1^i & a_2^i & \dots & a_{m-h}^i \\ \vdots & & & \\ a_1^{m-h} & a_2^{m-h} & \dots & a_{m-h}^{m-h} \end{vmatrix}$$

and therefore is equal to the unknown $v_{1i}^{(m-h)}$ in Eq. 2.43. In general, then,

$$v_{ji}^{(m-h)} = v_{ji}^{(m)} + \sum_{p=m-h+1}^{m} \Delta_{jp}^{(m-h+1)} v_{pi}^{(m)} \qquad \begin{matrix} i = 1, 2, \dots, m-h \\ j = 0, 1, \dots, m-h. \end{matrix} \tag{2.44}$$

This expression is identical to Eq. 1.42.

Chapter 3

Rapid Injection of Tracers

Two methods of administration of labeled hormones are commonly used in tracer experiments: infusion at a constant rate or rapid injection of a single dose. The infusion method is discussed in Chapter 2 where formulas to calculate rates are derived by solving linear algebraic equations that relate these rates to specific activities measured at the isotopic steady state. The second method is now considered in which a bolus of tracers is administered and specific activities are measured at various intervals following the injection.

The plotting of specific activities versus the time of sampling yields a curve that demonstrates the experimental information available for the calculation of rates. Almost all rates and most of the other kinetic parameters (transfer factors, contribution factors) described in Chapter 1 can be calculated from areas under specific activity curves if two or more tracers are used. Other interesting calculations, described in this chapter and in Chapter 5, are based on the knowledge of the specific activity functions, i.e., the "shape" of the curve.

A. Calculations Based on Areas under Specific Activity Curves

Production Rates (PR) and Transfer Factors (ϱ)

If a dose S_1^* (dpm) of the tracer is administered into pool 1, a fraction $\varrho_{1i} S_1^*$ reaches pool i (Eq. 2.16). Since PR_1 is defined as the rate of irreversible removal of material from pool i at the steady state, the following expression describes the identity between the net amount of isotope reaching and leaving pool i:

$$\varrho_{1i} S_1^* = \int_0^\infty PR_i \, \sigma_i^1(t) \, dt$$

where $\sigma_i^1(t)$ is the specific activity of the compound in pool i at time t after the administration of the tracer. To simplify the notation, $\sigma_i^1(t)$ is written as σ_i^1. Therefore,

$$PR_i = \frac{\varrho_{1i} S_1^*}{\int_0^\infty \sigma_i^1 dt} . \tag{3.1}$$

In particular, when $i=1$,

$$PR_1 = \frac{S_1^*}{\int_0^\infty \sigma_1^1\,dt} \tag{3.2}$$

and, in general,

$$PR_i = \frac{S_i^*}{\int_0^\infty \sigma_i^i\,dt}. \tag{3.3}$$

It follows from Eqs. 3.1 and 3.3 that

$$\varrho_{1i} = PR_i \frac{\int_0^\infty \sigma_i^1\,dt}{S_1^*} = \frac{S_i^*}{S_1^*} \frac{\int_0^\infty \sigma_i^1\,dt}{\int_0^\infty \sigma_i^i\,dt} \tag{3.4}$$

in general,

$$\varrho_{ji} = \frac{S_i^*}{S_j^*} \frac{\int_0^\infty \sigma_i^j\,dt}{\int_0^\infty \sigma_i^i\,dt} \qquad i, j = 1, 2, \ldots, m. \tag{3.5}$$

In summary, the production rate of a pool can be estimated by injecting a tracer into the pool and measuring the total area under the curve obtained by plotting the specific activity of the compound in the pool against the time elapsed after the injection (Eq. 3.3). The transfer factor from one pool to another can be estimated by injecting tracers labeled with different isotopes into both the donor and recipient pools and by measuring the total areas under the specific activity curves of the compound in the recipient pool (Eq. 3.5). Thus,

$$PR_i = \frac{S_i^*}{\int_0^\infty \sigma_i^i\,dt} = \frac{\text{Dose injected}}{\text{Total area under } \sigma_i^i(t)}$$

and

$$\varrho_{ji} = \frac{S_i^*}{S_j^*} \frac{\int_0^\infty \sigma_i^j\,dt}{\int_0^\infty \sigma_i^i\,dt} = \frac{S_i^*}{S_j^*} \frac{\text{Total area under } \sigma_i^j(t)}{\text{Total area under } \sigma_i^i(t)}$$

These results are illustrated in Fig. 3.1.

Rates

As discussed in Chapter 1, all v's and contribution factors (Δ's) in a steady-state system of any number of primary pools embedded in a multicompartmental network can be estimated if the pertinent PR and ϱ values are known

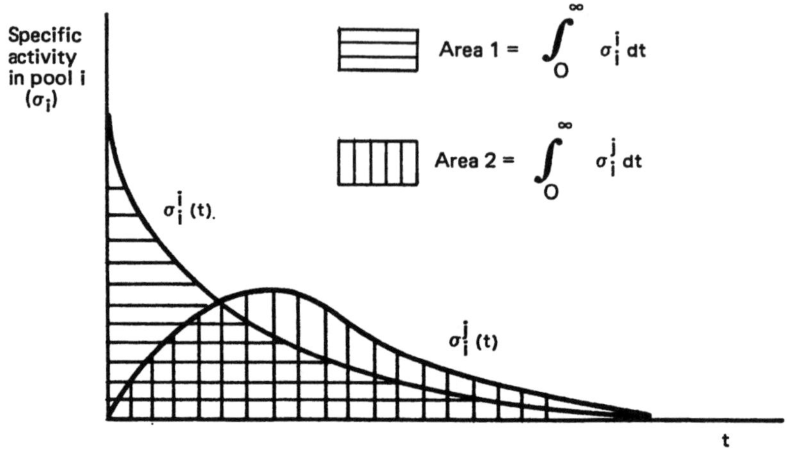

$$PR_i = \frac{S_i^*}{\text{Area 1}} = \frac{\rho_{ji} S_j^*}{\text{Area 2}}$$

$$\rho_{ji} = \frac{S_i^*}{S_j^*} \frac{\text{Area 2}}{\text{Area 1}}$$

Fig. 3.1. Calculation of production rates and transfer factors from areas under specific activity curves. The two curves represent the specific activities of the compound in pool i with respect to the isotope introduced into pool i $[\sigma_i^i(t)]$ or into pool j $[\sigma_i^j(t)]$, as a function of time following the injection of the tracers. Pools i and j are embedded in a multicompartmental system. S_i^* and S_j^* denote the doses (dpm) of the tracer, PR_1 is the production rate of the compound in pool i (nmole/hr) and ϱ_{ji} is the transfer factor corresponding to the movement of material from pool j to pool i. Area $1 = \int_0^\infty \sigma_i^i\,dt$ and area $2 = \int_0^\infty \sigma_i^j\,dt$ (see Eqs. 3.1 – 3.5).

(Eqs. 1.32 and 2.38; Tables 1.1 and 1.3). Therefore, Eqs. 3.3 and 3.5 provide solutions for the calculation of rates from isotopic data obtained from experiments in which single doses of the tracers are administered rapidly.

Estimation of the rates of recycle, w_i, requires knowledge of the "shape" of the specific activity curve.

B. Relationships Between Data Obtained by Rapid Injection and by Constant Infusion of a Tracer

It follows from Eqs. 2.1 and 3.3 that

$$PR_i = \frac{I_i^*}{a_i^i} = \frac{S_i^*}{\int_0^\infty \sigma_i^i\,dt} \,.$$

A more general expression is obtained from Eqs. 2.17 and 3.1:

$$\mathrm{PR}_1 = \frac{\varrho_{ji}\,\mathrm{I}_j^*}{\mathrm{a}_i^j} = \frac{\varrho_{ji}\,\mathrm{S}_j^*}{\int_0^\infty \sigma_i^j\,\mathrm{dt}}\,. \tag{3.6}$$

Therefore,

$$\frac{\mathrm{a}_i^j}{\mathrm{I}_j^*} = \frac{\int_0^\infty \sigma_i^j\,\mathrm{dt}}{\mathrm{S}^*} \qquad \mathrm{i, j} = 1, 2, \ldots, \mathrm{m}\,. \tag{3.7}$$

It is worth noting that Eq. 3.7 is based on integral values of the specific activities from time 0 to infinity and therefore is valid regardless of the length of time required to administer the dose S_j^* and regardless of discontinuities or fluctuations in the rate of administration.

The relationship shown in Eq. 3.7 has been explicitly described by MEIER and ZIERLER (1954) in reference to experimental designs to measure blood-flow proposed by STEWART (1897) and HAMILTON *et al.* (1932).

All formulas listed in Chapter 2 (e.g., Table 2.1) can be applied to data obtained from rapid injection tracer experiments by replacing steady-state specific activities (a's) by areas under specific activity curves (area $= \int_0^\infty \sigma\,\mathrm{dt}$) and infusion rates ($I^*$'s) by injected doses ($S^*$'s). Table 3.1 shows the formulas derived in this manner from the equations listed in Table 2.1.

The relationship between specific activities obtained during infusion of tracers at a constant rate or after instantaneous injection of a single dose is not limited to the situation described by Eq. 3.7 but also applies to any value of time, i.e. (GURPIDE *et al.*, 1963 b).

$$\frac{\mathrm{a}_{i,\tau}^j}{\mathrm{I}_j^*} = \frac{\int_0^\tau \sigma_i^j\,\mathrm{dt}}{\mathrm{S}_j^*} \tag{3.8}$$

where $a_{i,\tau}^j$ denotes the specific activity of the compound in pool i at time τ after starting the infusion of the tracer into pool j.

The relationship described by Eq. 3.8, of which Eq. 3.7 is a special case ($\tau \to \infty$), can be justified graphically, in the following manner.

Consider that the tracer is infused into pool 1 at the constant rate I_1^*. The infusion can be simulated by a succession of single injections at intervals Δt, each injection consisting of an amount of tracer equal to $I_1^*\Delta t$. The smaller the interval Δt, the more nearly equivalent the two procedures. As shown in Fig. 3.2, the amount $x_{1,\tau}^1$ of isotope in the pool at any time τ after the start of the infusion equals the sum of the contributions of each of the equivalent successive injections, i.e., it equals the heights of the ends of each of the curves at $t = \tau$. Since the amount of isotope present at time t following a rapid injection of a dose S_1^* can be expressed by

$$\mathrm{y}_1^1 = \mathrm{S}_1^*\,\mathrm{f}\,(\mathrm{t}) \tag{3.9}$$

Table 3.1. Formulas used to calculate rates of formation, transfer, and removal in a two-primary-pool subsystem (Fig. 1.2) from areas under specific curves [a]

$$PR_1 = \frac{S_1^*}{\int_0^\infty \sigma_1^1 \, dt} \qquad PR_2 = \frac{S_2^*}{\int_0^\infty \sigma_2^2 \, dt}$$

$$\varrho_{12} = \frac{S_2^*}{S_1^*} \frac{\int_0^\infty \sigma_2^1 \, dt}{\int_0^\infty \sigma_2^2 \, dt} \qquad \varrho_{21} = \frac{S_1^*}{S_2^*} \frac{\int_0^\infty \sigma_1^2 \, dt}{\int_0^\infty \sigma_1^1 \, dt}$$

$$\Delta_{12} = \frac{\int_0^\infty \sigma_2^1 \, dt}{\int_0^\infty \sigma_1^1 \, dt} \qquad \Delta_{21} = \frac{\int_0^\infty \sigma_1^2 \, dt}{\int_0^\infty \sigma_2^2 \, dt}$$

$$D = \int_0^\infty \sigma_1^1 \, dt \int_0^\infty \sigma_2^2 \, dt - \int_0^\infty \sigma_2^1 \, dt \int_0^\infty \sigma_1^2 \, dt$$

$$v_{01} = S_1^* \frac{\int_0^\infty \sigma_2^2 \, dt - \int_0^\infty \sigma_1^2 \, dt}{D} \qquad v_{02} = S_2^* \frac{\int_0^\infty \sigma_1^1 \, dt - \int_0^\infty \sigma_2^1 \, dt}{D}$$

$$v_{12} = S_2^* \frac{\int_0^\infty \sigma_2^1 \, dt}{D} \qquad v_{21} = S_1^* \frac{\int_0^\infty \sigma_1^2 \, dt}{D}$$

$$v_{10} = \frac{S_1^* \int_0^\infty \sigma_2^2 \, dt - S_2^* \int_0^\infty \sigma_2^1 \, dt}{D} \qquad v_{20} = \frac{S_2^* \int_0^\infty \sigma_1^1 \, dt - S_1^* \int_0^\infty \sigma_1^2 \, dt}{D}$$

[a] S_1^* and S_2^* denote the doses of tracers (dpm) injected into pools 1 and 2, respectively. The areas under the curves resulting from plotting specific activities (σ) versus time elapsed after the injection are given by the expressions $\int_0^\infty \sigma \, dt$. For instance, $\int_0^\infty \sigma_2^1 dt$ indicates the total area under the curve corresponding to the specific activity of the compound in pool 2 with respect to the isotope introduced with the tracer administered into pool 1.

it follows that each hypothetical injection contributes an amount of isotope given by the equation

$$y_1^1 = I_1^* \Delta t \, f(\tau - t') .$$

Note that I_1^* is the dose administered in the interval Δt and that $\tau - t'$ is the time elapsed between the time of injection, denoted by t', and the time of measurement. Therefore,

$$x_{1,\tau}^1 = \sum_{t'} I_1^* \Delta t \, f(\tau - t') = I_1^* \Delta t \sum_{t'} f(\tau - t') ,$$

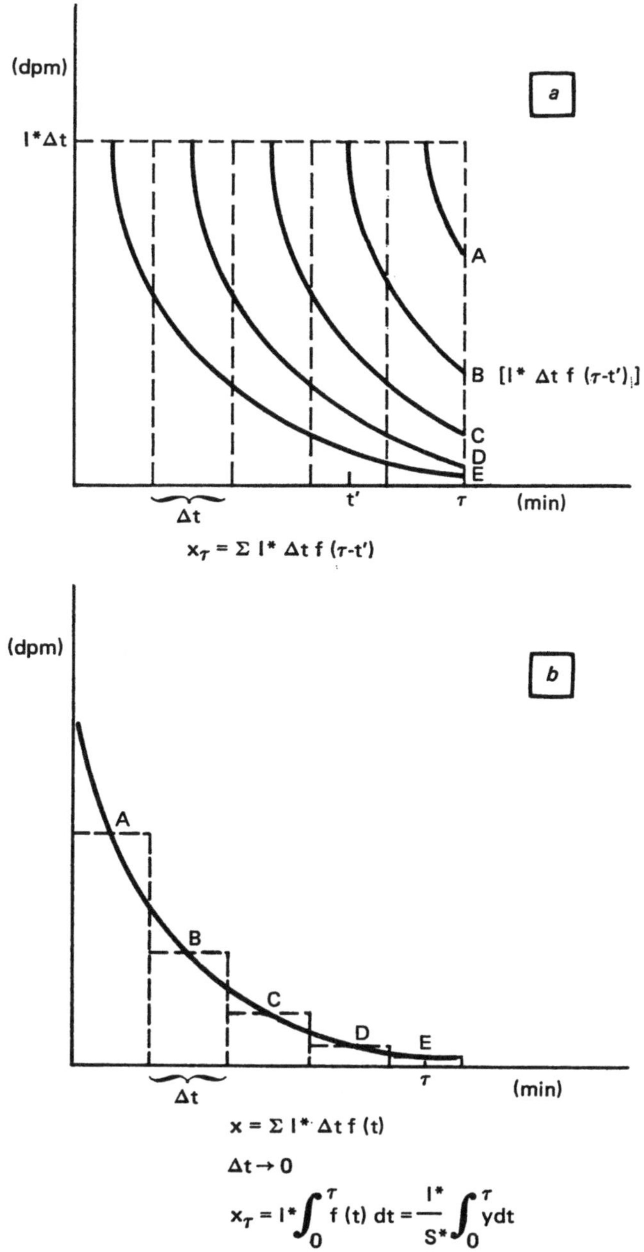

Fig. 3.2. Graphic derivation of the equivalence between single injection and constant infusion of a tracer. (a) Points A, B, ..., H represent the amount of isotope in a pool at time τ resulting from doses $I_1^* \Delta t$ (dpm) administered into the pool at various times. I_1^* denotes a rate of infusion (dpm/min). The amount of radioactivity in the pool at time τ equals the sum of the quantities A, B, ..., H, i.e., $x_{1,\tau}^1 = \sum_{t'} I_1^* \Delta t\, f(\tau - t')$. (b) If each of the segments of curves in Fig. 3.2 a are moved to a common origin (t = 0), it is apparent that $x_{1,\tau}^1 = \sum_t I_1^*\, f(t)\, \Delta t$

where the symbol $\Sigma_{t'}$ indicates a summation of all the terms, which differ in the value of t'.

If each of the segments of curve in Fig. 3.2 a are moved to a common origin, t = 0, they become superimposed; the lengths, however, remain different. As seen in Fig. 3.2 b, in which all curves in Fig. 3.2 a were brought to the same origin,

$$x_{1,\tau}^{1} = \sum_{t} I_1^* \, \Delta t \, f(t) = I_1^* \sum_{t} f(t) \, \Delta t \, .$$

It can now be considered that the interval Δt tends to zero. Then,

$$x_{1,\tau}^{1} = I_1^* \int_0^{\tau} f(t) \, dt \qquad (3.10)$$

and, replacing f (t) by y_1^1/S_1^* (Eq. 3.9),

$$x_{1,\tau}^{1} = \frac{I_1^*}{S_1^*} \int_0^{\tau} y_1^1 \, dt \, . \qquad (3.11)$$

If each side of this equality is divided by the amount of the compound in the pool, the isotope contents are converted to specific activities, i.e.,

$$a_{1,\tau}^{1} = \frac{I_1^*}{S_1^*} \int_0^{\tau} \sigma_1^1 \, dt$$

or

$$\frac{a_{1,\tau}^{1}}{I_1^*} = \frac{\int_0^{\tau} \sigma_1^1 \, dt}{S_1^*} \, . \qquad (3.12)$$

Figure 3.3 illustrates the equivalence between the data obtained by the two different methods of administering the tracer. It is apparent that the area under the curve obtained by plotting the specific activity of a pool per unit of radioactivity injected (dpm per μmole of compound per dpm injected) versus time, can be used to calculate the specific activity (dpm per μmole per dpm infused per hour) that would result if the tracer were infused rather than injected.

The procedure used to derive these equations is also applicable to cases when the pool sampled is not the one into which the tracer is introduced. The general form of Eq. 3.12 is Eq. 3.8.

Dividing Eq. 3.8 by Eq. 3.7, it follows that

$$\frac{a_{1,\tau}^{j}}{a_1^j} = \frac{\int_0^{\tau} \sigma_1^j \, dt}{\int_0^{\infty} \sigma_1^j \, dt} = \frac{\text{Area from 0 to } \tau}{\text{Total area}} \, . \qquad (3.13)$$

This expression indicates that interrupting the infusion at a time τ or collecting samples only until time τ after an injection results in data equally short

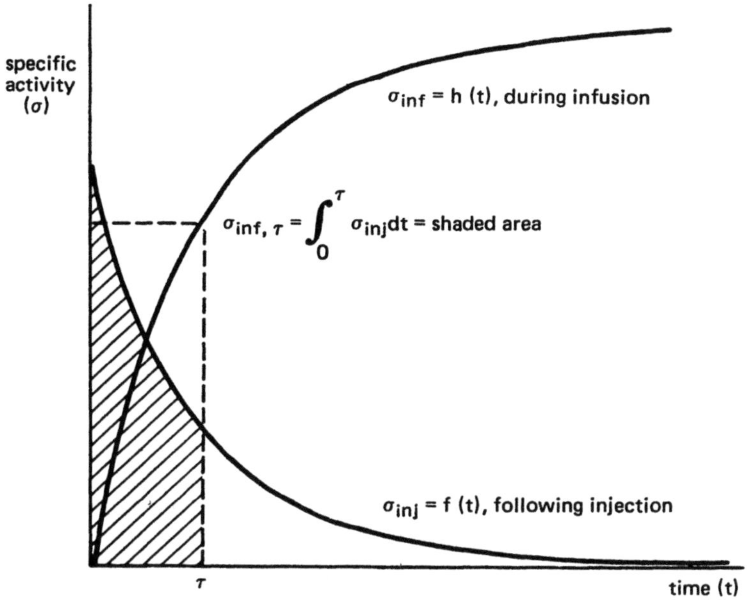

Fig. 3.3. Equivalence between specific activities obtained during single injection and constant infusion of a tracer. Typical curves corresponding to the specific activity in a pool of a system in the steady state following administration of the tracer into the pool by rapid injection (σ_{inj}) or by infusion at a constant rate (σ_{inf}). According to Eq. 3.12, $(\sigma_{inf})_\tau = \int_0^\tau \sigma_{inj}\, dt$ if the concentrations are expressed as percentages of the injected dose or the infusion rate

of the final ($t = \infty$) values. Therefore, the time of infusion required for an acceptable approximation to the isotopic steady state can be estimated by examination of the specific activity curve in a single-shot experiment.

By differentiation of Eq. 3.8, we obtain the useful equation

$$\frac{1}{I_j^*}\frac{da_{l,t}^j}{dt} = \frac{\sigma_l^j}{S_j^*}. \tag{3.14}$$

At $t = \infty$, $\sigma_l^j = 0$ and $da_{l,t}^j/dt = 0$, i.e., $a_{l,t}^j$ becomes constant.

C. Calculations Based on the Shape of Specific Activity Curves

The calculations described in the preceding sections are based upon measurements of the areas under specific activity curves, and are applicable to pools embedded in multicompartmental systems of an undetermined number of pools. Further information can be obtained from the actual function of time of the specific activities.

The data provided by the "shape" of the curves can be used in two general groups of calculations: (1) those based on the acceptance that the pools are part of a large and undefined number of interconnected pools, and (2) those based on the postulation of models consisting of a known, small number of pools with well-defined interconnections. The first group of calculations is considered below. The calculations based on models, useful in particular situations in spite of their limited applicability, are described in Chapter 4.

Total Number of Pools in the System Undetermined

The isotope content in any pool of a compartmentalized system in which the movement of isotope follows first-order kinetics, is characterized by multi-exponential functions (Chapter 4). Thus, if the system is at the steady state with respect to the unlabeled molecules,

$$\sigma = \sum_i D_i\, e^{-\alpha_i t}\,. \tag{3.15}$$

Experimentally, data obtained from *in vivo* tracer experiments can be adequately fit by curves representing sums of a few exponential terms. In the following discussion it is assumed that the specific activity functions are accurate, from $t=0$ to $t=\infty$. Critical comments on practical aspects related to the experimental determination and accuracy of these functions are presented in Chapter 5.

Pool Size

The amount M_1 of the compound present in the pool at the time of administration of a dose S_1^* of the tracer can be calculated from the sum of the coefficients since, when $t=0$ (Note 3.1),

$$(\sigma_1^1)_{t=0} = \frac{S_1^*}{M_1} = \sum D_i\,. \tag{3.16}$$

Note 3.1

Rules for the differentiation and integration of exponential functions are found in mathematical tables (e.g., Selby, 1972). These rules, when applied to the sum of two exponential terms with constant coefficients, e.g.,

$$\sigma = D_1\, e^{-\alpha_1 t} + D_2\, e^{-\alpha_2 t} = \frac{D_1}{e^{\alpha_1 t}} + \frac{D_2}{e^{\alpha_2 t}}$$

yield the following results:

$$(\sigma)_{t=0} = D_1 + D_2 \qquad \text{since } e^{\circ} = 1$$

$$(\sigma)_{t=\infty} = 0$$

$$\frac{d\sigma}{dt} = -\alpha_1 D_1\, e^{-\alpha_1 t} - \alpha_2 D_2\, e^{-\alpha_2 t}$$

$$\left(\frac{d\sigma}{dt}\right)_{t=0} = -\alpha_1 D_1 - \alpha_2 D_2$$

$$\int \sigma\, dt = \frac{D_1}{-\alpha_1} e^{-\alpha_1 t} + \frac{D_2}{-\alpha_2} e^{-\alpha_2 t} + \text{constant}$$

$$\int_0^\infty \sigma\, dt = \left[\frac{D_1}{-\alpha_1} e^{-\alpha_1 t} + \frac{D_2}{-\alpha_2} e^{-\alpha_2 t}\right]_0^\infty = \frac{D_1}{\alpha_1} + \frac{D_2}{\alpha_2}.$$

Rates of Exit

The rate of change in the amount of isotope in a pool is equal to the difference between the rates at which the isotope enters and leaves the pool. As indicated in the Introduction to this monograph, since the rate at which the isotope leaves the pool is a first-order process in a system at the steady state with respect to the endogenous compounds (Note 3.2), then

$$\frac{dy^*}{dt} = -k_T y^* + f(t) \tag{3.17}$$

where k_T is the rate constant of total removal of the tracer, y^* is the amount of isotope, and $f(t)$ is the rate of reentry of the isotope into the pool. The rate constant is defined as the ratio of the rate at which the compound leaves the pool (v_T) and the amount of the compound present in the pool (M). If the tracer is administered into pool 1 at $t = 0$,

$$\left(\frac{dy_1^1}{dt}\right)_{t=0} = -k_{T,1}(y_1^1)_{t=0} \tag{3.18}$$

since at the moment of injection all the isotope is in pool 1 and none is returning ($f(0) = 0$). If both sides of Eq. 3.18 are divided by M_1, the total amount of the compound in the pool, a similar expression in terms of specific activities ($\sigma_1^1 = y_1^1/M_1$) is obtained, i.e.,

$$\left(\frac{d\sigma_1^1}{dt}\right)_{t=0} = -k_{T,1}(\sigma_1^1)_{t=0} \tag{3.19}$$

or

$$k_{T,1} = \frac{(d\sigma_1^1/dt)_{t=0}}{(\sigma_1^1)_{t=0}}. \tag{3.20}$$

By calculating the first derivative of Eq. 3.15, it follows that

$$\frac{d\sigma_1^1}{dt} = -\sum \alpha_i D_i e^{-\alpha_i t}$$

and, at $t = 0$,

$$\left(\frac{d\sigma_1^1}{dt}\right)_{t=0} = -\sum D_i \alpha_i. \tag{3.21}$$

From Eqs. 3.16, 3.19, and 3.21, it follows that

$$k_{T,1} = \frac{\sum D_i \alpha_i}{\sum D_i} . \tag{3.22}$$

D's and α's can be estimated from the experimental curve using methods described in the last section of this chapter. Since the rate constant $k_{T,1}$ equals the sum of all rates of removal of the compound from pool 1 ($v_{T,1} = v_{10} + w_1$ in Fig. 1.1) divided by the amount of compound in the pool (M_1), it follows from Eq. 3.16 and 3.22 that

$$v_{T,1} = k_{T,1} M_1 = \frac{(d\sigma_1^1/dt)_{t=0} S_1^*}{(\sigma_1^1)^2_{t=0}} = S_1^* \frac{\sum D_i \alpha_i}{(\sum D_i)^2} . \tag{3.23}$$

Note 3.2

According to concepts of chemical kinetics, the rate at which the tracee present in a pool at concentration c_1 participates in a specific process depends upon the concentration of all reagents involved, as indicated by the following equation:

$$r = k (c_1)^l \prod_j (c_j)^m$$

where c_j is the concentration of reagent j, and l and m are the number of molecules of tracee and of reagent j participating in the reaction.

If a tracer is introduced into the pool, the total rate of the forward reaction equals the sum of the rates of reaction of tracer and tracee, i.e.,

$$r + r^* = k (c_1 + c_1^*)^l \prod_j (c_j)^m$$

or, by binominal expansion (SELBY, 1972),

$$r + r^* = k \left[c_1^l + l\, c_1^{l-1} c_1^* + \frac{l(l-1)}{2} c_1^{l-2} c_1^{*2} + \cdots \right] \prod_j (c_j)^m .$$

When $l = 1$ or when c_1^* is small enough with respect to c_1 so that the terms of the binominal expansion involving powers of c_1^* can be neglected, then

$$r + r^* = k (c_1^l + l\, c_1^{l-1} c_1^*) \prod_j (c_j)^m$$

$$= r + k\, l\, c_1^{l-1} c_1^* \prod_j (c_j)^m .$$

Therefore, if the system is at the steady state and the concentrations of endogenous compounds remain constant, it follows that

$$r^* = K c_1^*$$

This expression indicates that under these conditions, the reaction is pseudo-first-order with respect to the tracer.

Rates of Reentry, Fractional Loss

The rate or reentry of the compound into the pool (w_1 in Fig. 1.1) can be calculated from isotopic data since $w_1 = v_{T,1} - v_{10}$ and expressions to estimate $v_{T,1}$ and v_{10} (PR_1 at the steady state) are also available (Eqs. 2.23 and 3.2, respectively).

The rate w_1 in a system of two pools embedded in a multicompartmental network (Fig. 1.2) can also be calculated since, in this case, $w_1^{(2)} = v_{T,1} - v_{10}^{(2)} - v_{12}^{(2)}$, and formulas to estimate $v_{T,1}$, $v_{10}^{(2)}$, and $v_{12}^{(2)}$ from isotopic data are given (Eq. 3.27 and Table 3.1).

In general, for an m-primary-pool system

$$w_i^{(m)} = v_{T,i} + v_{ii}^{(m)} . \qquad (3.24)$$

It is also possible to estimate the fraction (φ) of the total amount of a compound leaving the pool that is lost irreversibly:

$$\varphi_1 = \frac{PR_1}{v_{T,1}} . \qquad (3.25)$$

If the experimental specific activity curve is described by Eq. 3.15 then, from Eq. 3.2, $PR_1 = S_1^*/\int_0^\infty \sigma_1^1 dt = S_1^*/\Sigma\,(D_i/\alpha_i)$. It follows from this expression and from Eq. 3.23 that

$$\varphi_1 = \frac{(\sum D_i)^2}{\sum D_i\,\alpha_i \sum (D_i/\alpha_i)} . \qquad (3.26)$$

"Turnover Times"

Several parameters can be included under the denomination of "turnover time," which, therefore, is not an accurately defined term, although it is currently used (BROWNELL *et al.*, 1968). Thus, when a compound is recycled through a pool, at least two time parameters must be recognized: the average total time ($\overline{T}$) that a molecule spends in the pool including all recycles, and the average time ($\overline{t}$) spent by the molecule in one passage through the pool.

Reentry of a compound into a pool (e.g., reentry of a hormone into the systemic circulation) must be considered whenever more than one exponential term is required to describe the decline with time of the specific activity of the compound in the pool into which the tracer is injected.

Recent publications (RESCIGNO and GURPIDE, 1973; RESCIGNO, 1973) defined various time parameters and presented formulas for their calculation from data obtained by performing tracer experiments. Some of these results are now described.

Mean Residence Time ($\overline{T}$)

The first of the parameters listed above is called the residence time of the compound in the pool, and is denoted by the symbol $\overline{T}$. If each molecule stays in the pool an average time $\overline{T}$, an amount equal to the content of the pool (M) would be irreversible removed from the pool in a period of time $\overline{T}$. Therefore, since PR_1 equals the rate of irreversible removal of the compound from the pool, it follows that $PR = M/\overline{T}$. By replacing PR and M by the corresponding expressions in experimental terms (Eqs. 3.2 and 3.16) the following result is obtained:

$$\overline{T}_1 = \frac{M}{PR_1} = \frac{1}{(\sigma_1^1)_{t=0}} \int_0^\infty \sigma_1^1 dt = \frac{\sum (D_i/\alpha_i)}{\sum D_i} . \tag{3.27}$$

Mean Transit Time ($\overline{t}$)

The second time parameter, called the mean transit time of the compound through the pool and denoted by $\overline{t}$, is the average period of time during which the molecules of the compound stay during one passage through the pool. Since, by this definition, an amount equal to M will leave the pool in a period of time $\overline{t}$, it follows that $v_T = M/\overline{t}$ and, from Eqs. 3.16 and 3.23, that

$$\overline{t}_1 = \frac{M_1}{v_{T,1}} = \frac{\sum D_i}{\sum D_i \alpha_i} . \tag{3.28}$$

Therefore, the ratio between $\overline{t}_1$ and $\overline{T}_1$ equals the fraction φ_1 (Eq. 3.25), i.e.,

$$\frac{\overline{t}_1}{\overline{T}_1} = \frac{M_1}{v_{T,1}} \frac{PR_1}{M_1} = \frac{PR_1}{v_{T,1}} = \varphi_1 . \tag{3.29}$$

Mean Number of Cycles (ν)

If ν is defined as the average number of recycles of molecules through a pool, when $\nu_1 = 0$ there is no recycle and $\overline{T}_1 = \overline{t}_1$. When $\nu_1 = 1$, $\overline{T}_1 = 2\,\overline{t}_1$. Then,

$$\frac{\overline{T}_1}{\overline{t}_1} = \nu_1 + 1$$

or, from Eq. 3.29,

$$\nu_1 = \frac{1}{\varphi_1} - 1 . \tag{3.30}$$

Mean Recycling Time ($\overline{t}_{ii}$)

The mean recycling time, $\overline{t}_{ii}$, of a compound around pool i is defined as the average interval of time spent by a molecule of the compound from its exit from the pool to its next reentry into it.

An expression to estimate $\bar{t}_{11}$ from tracer data can be derived as follows. The fraction

$$\frac{PR_1\,\sigma_1^1}{\int_0^\infty PR_1\,\sigma_1^1\,dt} = \frac{\sigma_1^1}{\int_0^\infty \sigma_1^1\,dt}$$

represents the fractional rate at which the labeled hormone injected into pool 1 is removed irreversibly from the pool at time t. Therefore,

$$P_1^1 = \int_0^\infty t \left(\frac{\sigma_1^1}{\int_0^\infty \sigma_1^1\,dt} \right) dt = \frac{\int_0^\infty t\,\sigma_1^1\,dt}{\int_0^\infty \sigma_1^1\,dt} \tag{3.31}$$

is the expected time a molecule of the tracer will leave the pool irreversibly. Note that P_1^1 is larger than $\bar{T}_1$, as $\bar{T}_1$ includes only the average time spent by a molecule in pool 1, whereas P_1^1 refers to the total average time of stay of the labeled fraction of the molecule in the whole system during recycle.

Since $P_1^1 - \bar{T}_1$ is the average interval of time spent by the labeled fractions of a molecule of the compound outside pool 1 before being removed irreversibly from the pool, then the mean recycle time is given by the expression

$$\bar{t}_{11} = \frac{P_1^1 - \bar{T}_1}{\nu_1}. \tag{3.32}$$

Mean Transfer Time ($\bar{t}_{ij}$)

The average time required for the transfer of a molecule (or a labeled portion of it) from pool i to pool j is called the mean transfer time between these pools and is denoted by the symbol $\bar{t}_{ij}$. This time corresponds to the interval spent in other pools during the transfer. If there were no intermediate pools involved, the mean transfer time $\bar{t}_{ij}$ would be zero. If the transfer were to occur through only one intermediate pool (pool k), then $\bar{t}_{ij} = \bar{t}_k$. The discussion to follow leads to an expression from which $\bar{t}_{ij}$ can be estimated from isotopic data.

The same reasoning followed to define the time P_1^1 (Eq. 3.31) is applied to define three other similar parameters:

$$P_2^2 = \frac{\int_0^\infty t\,\sigma_2^2 dt}{\int_0^\infty \sigma_2^2 dt}, \tag{3.33}$$

the expected time at which a labeled molecule introduced into pool 2 at t = 0 leaves the system forever from pool 2;

$$P_1^2 = \frac{\int_0^\infty t\,\sigma_1^2\,dt}{\int_0^\infty \sigma_1^2 dt}, \tag{3.34}$$

the expected time at which a molecule introduced into pool 2 at $t = 0$ leaves the system forever from pool 1; and

$$P_2^1 = \frac{\int_0^\infty t\, \sigma_2^1 dt}{\int_0^\infty \sigma_2^1 dt}, \qquad (3.35)$$

the expected time at which a molecule introduced into pool 1 at $t = 0$ leaves the system forever from pool 2.

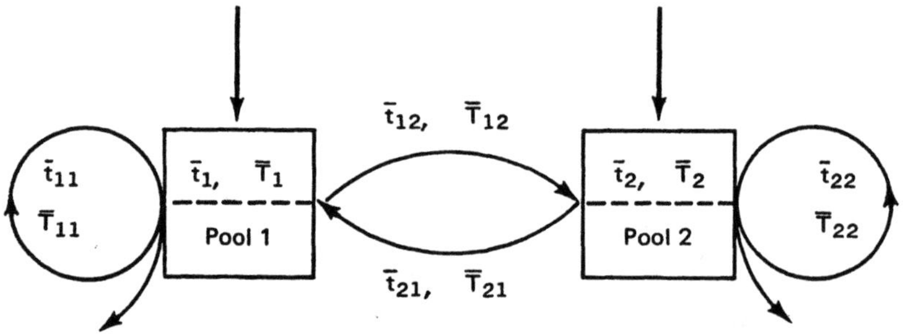

Fig. 3.4. Times of transit, residence, transfer, and recycle in a two-pool subsystem. Represented in the figure are the mean transit times of a compound through a pool in a single passage ($\bar{t}_1$ and $\bar{t}_2$), the mean time of residence of a compound in a pool during all recycles ($\bar{T}_1$ and $\bar{T}_2$), the average time of transfer of material from one pool to another in a single passage ($\bar{t}_{12}$ and $\bar{t}_{21}$) or during all recycles ($\bar{T}_{12}$ and $\bar{T}_{21}$), and the mean time of recycle around either pools 1 or 2 without passing through the other pool ($\bar{t}_{11}$ and $\bar{t}_{22}$ for one cycle, $\bar{T}_{11}$ and $\bar{T}_{22}$ when all cycles are included). Formulas to calculate these parameters from isotope data are given (Eqs. 3.27, 3.28, 3.37 to 3.40)

These experimentally determinable values can be related to the rates shown in Fig. 1.2 (which can be estimated from the same isotopic data) and to the mean times $\bar{t}_{11}$, $\bar{t}_{22}$, $\bar{t}_{12}$, and $\bar{t}_{21}$ (Fig. 3.4) by four independent equations involving these four unknown times. Consider, for instance, the time P_1^1. P_1^1 can be resolved into the following terms: the average interval of time spent by a molecule in its first passage through pool 1 ($\bar{t}_1$), the average interval of time spent in one cycle around pool 1 ($\bar{t}_{11} + \bar{t}_1$) times the expected number of cycles around it (A), the average interval of time spent in the reversible transfer to pool 2 ($\bar{t}_{12} + \bar{t}_2 + \bar{t}_{21} + \bar{t}_1$) times the expected number of such transfers (B), and the average interval of time spent in one cycle around pool 2 ($\bar{t}_{22} + \bar{t}_2$) times the expected number of such cycles by molecules that will eventually return to pool 1 (C). The three other equations, corresponding to P_2^1, P_2^2, and P_1^2 are similarly derived and the following set of equations results:

$$\begin{aligned}
P_1^1 &= \bar{t}_1 + A(\bar{t}_{11} + \bar{t}_1) + B(\bar{t}_{12} + \bar{t}_2 + \bar{t}_{21} + \bar{t}_1) + C(\bar{t}_{22} + \bar{t}_2)\\
P_2^1 &= P_1^1 + (\bar{t}_{12} + \bar{t}_2) + G(\bar{t}_{22} + \bar{t}_2)\\
P_2^2 &= \bar{t}_2 + F(\bar{t}_{22} + \bar{t}_2) + B(\bar{t}_{12} + \bar{t}_2 + \bar{t}_{21} + \bar{t}_1) + E(\bar{t}_{11} + \bar{t}_1)\\
P_1^2 &= P_2^2 + (\bar{t}_{21} + \bar{t}_1) + H(\bar{t}_{11} + \bar{t}_1)\,.
\end{aligned} \qquad (3.36)$$

The coefficients A, B, C, E, F, G, and H can be expressed in terms of the rates appearing in Fig. 1.2, as follows:

$$A = \frac{w_1 (v_{20} + v_{21})}{(v_{10} + v_{12})(v_{20} + v_{21}) - v_{12} v_{21}}$$

$$B = \frac{v_{12} v_{21}}{(v_{10} + v_{12})(v_{20} + v_{21}) - v_{12} v_{21}}$$

$$C = \frac{w_2}{v_{20} + v_{21}} B$$

$$E = \frac{w_1}{v_{10} + v_{12}} B$$

$$F = \frac{w_2 (v_{10} + v_{12})}{(v_{10} + v_{12})(v_{20} + v_{21}) - v_{12} v_{21}}$$

$$G = \frac{w_2}{v_{20} + v_{21}}$$

$$H = \frac{w_1}{v_{10} + v_{12}} .$$

These relations can be derived by intuitive reasoning (RESCIGNO and GURPIDE, 1973) and verified by a more direct analysis (RESCIGNO, 1973).

Solving the system of Eq. 3.36 and replacing the values of the coefficients A to H by the expression given above, it follows that

$$\bar{t}_{11} = \left(\frac{v_{T,1}}{w_1} - 1\right)\left[P_1^1 - \frac{v_{12} v_{21}}{(v_{10} + v_{12})(v_{20} + v_{21}) - v_{12} v_{21}} (P_1^2 + P_2^1 - P_1^1 - P_2^2)\right] - \frac{v_{T,1}}{w_1} \bar{t}_1 \tag{3.37}$$

$$\bar{t}_{22} = \left(\frac{v_{T,2}}{w_2} - 1\right)\left[P_2^2 - \frac{v_{12} v_{21}}{(v_{10} + v_{12})(v_{20} + v_{21}) - v_{12} v_{21}} (P_1^2 + P_2^1 - P_1^1 - P_2^2)\right] - \frac{v_{T,2}}{w_2} \bar{t}_2 \tag{3.38}$$

$$\bar{t}_{12} = \frac{(v_{10} + v_{12})(v_{20} + v_{21})}{(v_{10} + v_{12})(v_{20} + v_{21}) - v_{12} v_{21}} (P_1^2 + P_2^1 - P_1^1 - P_2^2) - P_1^2 \tag{3.39}$$

$$\bar{t}_{21} = \frac{(v_{10} + v_{12})(v_{20} + v_{21})}{(v_{10} + v_{12})(v_{20} + v_{21}) - v_{12} v_{21}} (P_1^2 + P_2^1 - P_1^1 - P_2^2) - P_2^1 . \tag{3.40}$$

The measurement of mean residence, transit, recycle, and transfer times of circulating hormones is discussed in Chapter 5.

Two-Primary-Pool Subsystem

Table 3.2 summarizes the calculation of parameters of two-primary-pool subsystems from specific activity functions in the pools. This table complements Table 3.1, which shows calculation based on areas under specific activity curves.

Table 3.2. Parameters of a two-primary-pool subsystem (Figs. 1.2 and 3.4) which can be calculated if specific activity functions are obtained from both pools [a]

$M_1 = S_1^*/(\sigma_1^1)_0$	$M_2 = S_2^*/(\sigma_2^2)_0$
$k_{T,1} = (d\sigma_1^1/dt)_0/(\sigma_1^1)_0$	$k_{T,2} = (d\sigma_2^2/dt)_0/(\sigma_2^2)_0$
$v_{T,1} = k_{T,1}\, M_1$	$v_{T,2} = k_{T,2}\, M_2$
$w_1 = v_{T,1} - v_{12} - v_{10}$	$w_2 = v_{T,2} - v_{21} - v_{20}$
$\varphi_1 = PR_1/v_{T,1}$	$\varphi_2 = PR_2/v_{T,2}$
$\nu_1 = (1 - \varphi_1)/\varphi_1$	$\nu_2 = (1 - \varphi_2)\, \varphi_2$
$\bar{t}_1 = M_1/v_{T,1}$	$\bar{t}_2 = M_2/v_{T,2}$
$\bar{T}_1 = M_1/PR_1$	$\bar{T}_2 = M_2/PR_2$
$\bar{t}_{11}, \bar{t}_{22}, \bar{t}_{12}, \bar{t}_{21}$ (Eqs. 3.37 to 3.40)	
$k_{irr,1} = PR_1/M_1$	$k_{irr,2} = PR_2/M_2$
$k_{rev,1} = k_{T,1} - k_{irr,1}$	$k_{rev,1} = k_{T,2} - k_{irr,2}$

[a] In addition to all the parameters listed in Table 3.1.

D. Determination of Areas and Shapes of Specific Activity Curves

Graphic Methods to Measure Areas

As discussed in the first section of this chapter, production rates and transfer factors (PR's and ϱ's), and therefore rates and contribution factors (v's and Δ's), can be determined from values of areas under specific activity curves. These calculations are possible even if the analytical expression describing the shape of the curve is not known. Note that the area is known if the specific activity function is known (area $= \int_0^\infty \sigma\, dt$) but the converse is not true.

A few direct graphic methods can be used to estimate areas under curves. In one of them the area is measured with the aid of a planimeter, a mechanical instrument that provides a measure of the area encompassed by a closed curve traced by a movable arm. In another method, the data are plotted on graph paper and the number of squares under the curve are counted. In a simple extension of this method, the area under the curve is represented as a sum of trapezoids obtained by considering that successive data points are connected by straight lines. Then, as indicated by NORMAND and FORTIER (1970) who emphasized the advantages of this method,

$$\text{Area}\Big|_{t_1}^{t_n} = \frac{\sigma_{(1)}(t_2 - t_1)}{2} + \frac{\sigma_{(n)}(t_n - t_{n-1})}{2} + \sum_{i=2}^{n-1} \frac{\sigma_{(i)}(t_{i+1} - t_{i-1})}{2} \qquad (3.41)$$

where $\sigma_{(i)}$ is the specific activity at time t_i. The area can then be conveniently obtained not only graphically but also numerically, using a calculator. However, graphic inspection of the data may be necessary to estimate values for the intercepts with the σ and t axes (values of σ at $t=0$ and of t for $\sigma \cong 0$, respectively). Comments on the uncertainties involved in these extrapolations are presented in Chapter 5.

Other geometric constructions that can be used to evaluate areas from the experimental data have been illustrated by RESCIGNO and GURPIDE (1973).

Procedures to Determine Specific Activity Functions "Peel-off" Graphic Method

The well-known "peel-off" graphic method is frequently used to determine the function $\sigma = f(t)$ (ROBERTSON, 1957; SOLOMON, 1960). This method is applicable to curves that can be adequately described by a sum of a few exponential terms. Most specific activity curves fall in this category.

If the function is represented by Eq. 3.15, it is apparent that only the term with the smallest α_i will be significant as t grows large. Thus, if the terms are ordered so that $\alpha_1 > \alpha_2 > \cdots > \alpha_m$,

$$\sigma \cong D_m e^{-\alpha_m t} \qquad \text{as } t \to \infty \tag{3.42}$$

or

$$\ln \sigma \cong -\alpha_m t + \ln D_m \qquad \text{as } t \to \infty . \tag{3.43}$$

Therefore, a plot of $\ln \sigma$ versus time would result in a curve asymtotic to a straight line. The slope of the straight final "tail" of the curve gives the value of α_m and can be estimated easily by determining the time necessary to reduce the specific activity to half its value (a constant for a straight line). If such time is called a "half-life", $t_{1/2}$, then

$$\ln \sigma - \ln \frac{\sigma}{2} = \alpha_m t_{1/2} \quad \text{or} \quad \ln 2 = \alpha_m t_{1/2}$$

and

$$\alpha_m = \frac{\ln 2}{t_{1/2}} = \frac{0.693}{t_{1/2}} .$$

The intercept of the extrapolated straight lines with the ordinate axis ($t=0$) gives the value of the coefficient D_m. Knowing α_m and D_m it is then possible to calculate the value of the term $D_m e^{-\alpha_m t}$ for any t. Therefore, a new function,

$$\sigma - D_m e^{-\alpha_m t} = \sum_{i=1}^{m-1} D_i e^{-\alpha_i t}$$

can now be considered. As before, a semilogarithmic plot of this function is expected to yield a straight line as a tail, from which α_{m-1} and D_{m-1} can be estimated.

This procedure is continued until the function resulting from successive substitutions of the calculated terms from the experimental data yields a straight line when plotted semilogarithmically, i.e.,

$$\sigma - \sum_{i=2}^{m} D_i\, e^{\alpha_i t} = D_1\, e^{-\alpha_1 t}\,.$$

This method is further discussed in Chapter 5 where an example of calculations based on experimental data is given (Note 5.1). Usually no more than four exponential terms are necessary to account satisfactorily for the specific activity functions obtained from tracer data from biologic systems.

Once the function $\sigma = f(t)$ is determined, the total area under the specific activity curve can be determined analytically. Thus, if $\sigma = \Sigma\, D_i\, e^{-\alpha_i t}$, then $\int_0^\infty \sigma\, dt = \Sigma\, (D_i/\alpha_i)$ (Note 3.1).

A description of computer-implemented numerical methods for the determination of specific activity functions best fitting the experimental data is presented in the next section, written by R. J. BOGUMIL.

Numerical (Computer) Methods

Computer "optimization" programs are widely available to fit data by algebraic functions (e.g., polynomials, trigonometric functions, exponentials, etc.), as well as by kinetic models (e.g., the "rate constants" of an n compartment system). The requisite mathematical techniques are the subject of extensive literature (see, for example, WHITTAKER and ROBINSON, 1924; BELLMAN, 1957; BOX, 1960; BERMAN *et al.*, 1962; MARQUARDT, 1963; and HIMMELBLAU, 1972). The following description is intended to provide a brief introduction to this subject with the objective of making the computer procedure less mysterious and possibly motivating further interest on the reader's part. The fitting of sums of exponentials to data will be considered explicitly. The multiexponential fit then may be further analyzed by the methods of Chapter 4. It is perhaps appropriate to mention, at the outset, that digital computers can perform only very simple arithmetic and logical operations, e.g., addition, subtraction, multiplication, division, and the logical *and*, *or*, and *not*. To accomplish a more complex calculation, such as the above optimizations, it is necessary for the mathematician to find an (approximate) method that will reduce the analytic expression to a (possibly large) number of simple operations. Inasmuch as the computer time required for a single arithmetic operation is on the order of a few microseconds, it may be permissible to use millions or tens of millions but, for economic reasons, probably not billions of elementary operations. Such limitations, as well as limits on memory size, arithmetic precision, and other factors, are the constant concern of programmers and applied mathematicians. Nevertheless, modern algorithms are, in large measure, based on techniques developed by Newton, Legendre, Gauss, and others, and have been applied to problems in physics and biology for over a century by human computers.

Numerical techniques have numerous advantages over graphic procedures in terms of repeatability and accuracy. The relative advantages are particularly important in the analysis of multiexponential curves in which exponents differ by less than an order of magnitude, or in the presence of certain types of experimental (measurement) errors. This is not to claim that even the computer-implemented methods are particularly accurate under such circumstances but rather that, properly applied, the techniques provide a uniform and reasonably sound approach to the problem. By the graphic method, an investigator may be forced into tedious repetition of the procedure in order to determine the sensitivity of his result to differing treatment of, for example, a possible outlier or the number of exponential terms. Computer-implemented procedures can be designed to provide comparable information (i.e., on the sensitivity of parameter estimates to various sources of error) and such features are, in fact, standard with many of the available routines.

We wish to fit a function $\sigma(t) = f(t, D_1, D_2, \ldots, D_m, \alpha_1, \alpha_2, \ldots, \alpha_m)$ to the observation time series in a manner so as to minimize some measure of the deviation between the analytic curve (i.e., the function we chose to fit) and the data. We have already stipulated that this function is to be a sum of exponentials with (to simplify matters) real valued parameters, i.e., the function is $\sigma(t) = D_1 e^{-\alpha_1 t} + \cdots + D_m e^{-\alpha_m t}$, where we wish to find "best values" of the D's and α's. These values will be influenced by the decision on the number of terms to use, i.e., the value of m. The procedure frequently employed, under circumstances for which there is no *a priori* basis for choosing a specific value for m, is to first fit a single term $\sigma(t) = D e^{-\alpha t}$, determine by one of several possible criteria whether the result is adequate, then fit $\sigma(t) = D_1 e^{-\alpha_1 t} + D_2 e^{-\alpha_2 t}$ by repeating the entire process, etc. Note that the optimum values for D and α will not correspond identically with any of the best values D_1, D_2, α_1, and α_2. The curve-fitting process may be accomplished by application of one of a variety of algorithms described in detail in references cited previously. The principles involved are illustrated by the procedure outlined below.

We shall designate the difference between the analytic curve and the data (at each of k data points y_i, $i = 1, \ldots, k$) by the symbol δ_i. Thus,

$$\delta_i = f(t_i, D_1, D_2, \ldots, D_m, \alpha_1, \alpha_2, \ldots, \alpha_m) - y_i \qquad i = 1, \ldots, k \tag{3.44}$$

where y_i represents the i-th experimental value (at time t_i). Our objective is to minimize some function of the δ_i, e.g., the sum of absolute values, sum of squared values, weighted sum, etc. The most common, and analytically most tractable choice is to minimize the sum of the squared deviations, i.e., minimize $\sum_{i=1}^{k} \delta_i^2$ for a given m. A necessary (but not sufficient) condition for a function to have a local minimum is, of course, that the derivative of the function be zero at a point. Thus, we seek values of D_j and α_j to satisfy the relations

$$\frac{\partial}{\partial D_j} \sum_{i=1}^{k} \delta_i^2 = 0 \qquad j = 1, \ldots, m$$

$$\frac{\partial}{\partial \alpha_j} \sum_{i=1}^{k} \delta_i^2 = 0 \qquad j = 1, \ldots, m. \tag{3.45}$$

By simple algebraic manipulation (and the rules of differential calculus)

$$\frac{\partial}{\partial D_j} \sum_{i=1}^{k} \delta_i^2 = 2 \sum_{i=1}^{k} \delta_i \frac{\partial \delta_i}{\partial D_j} \qquad j = 1, \ldots, m$$

$$\frac{\partial}{\partial \alpha_j} \sum_{i=1}^{k} \delta_i^2 = 2 \sum_{i=1}^{k} \delta_i \frac{\partial \delta_i}{\partial \alpha_j} \qquad j = 1, \ldots, m$$

so that (from 3.45) an equivalent relation is

$$\sum_{i=1}^{k} \delta_i \frac{\partial \delta_i}{\partial D_j} = 0 \qquad j = 1, \ldots, m$$

$$\sum_{i=1}^{k} \delta_i \frac{\partial \delta_i}{\partial \alpha_j} = 0 \qquad j = 1, \ldots, m. \tag{3.46}$$

We now substitute the expression for δ_i (Eq. 3.44) into Eqs. 3.46

$$\sum_{i=1}^{k} [f(t_i, D_1, D_2, \ldots, D_m, \alpha_1, \alpha_2, \ldots, \alpha_m) - y_i] \times \frac{\partial f(t_i, D_1, D_2, \ldots, D_m, \alpha_1, \alpha_2, \ldots, \alpha_m)}{\partial D_j} = 0 \qquad j = 1, \ldots, m$$

$$\sum_{i=1}^{k} [f(t_i, D_1, D_2, \ldots, D_m, \alpha_1, \alpha_2, \ldots, \alpha_m) - y_i] \times \frac{\partial f(t_i, D_1, D_2, \ldots, D_m, \alpha_1, \alpha_2, \ldots, \alpha_m)}{\partial \alpha_j} = 0 \qquad j = 1, \ldots, m$$

or, more specifically for a multiexponential function,

$$\sum_{i=1}^{k} [D_1 e^{-\alpha_1 t_i} + D_2 e^{-\alpha_2 t_i} + \ldots + D_m e^{-\alpha_m t_i} - y_i] e^{-\alpha_j t_i} = 0 \qquad j = 1, \ldots, m$$

$$\sum_{i=1}^{k} [D_1 e^{-\alpha_1 t_i} + D_2 e^{-\alpha_2 t_i} + \ldots + D_m e^{-\alpha_m t_i} - y_i] t_i D_j e^{-\alpha_j t_i} = 0 \qquad j = 1, \ldots, m. \tag{3.47}$$

Our conjecture is that those values of the D's and α's that are approximate solutions to the above system of simultaneous algebraic equations yield a minimum sum of squares difference between the data and the analytic curve. Note that the equations to be solved possess 2 m unknowns (in 2 m equations). If these equations were linear and the coefficient matrix nonsingular, we could be confident, not only that a single solution exists, but that infallible methods for its calculation (subject to certain restrictions on dimension and other matters beyond the scope of this outline) are known. Unfortunately, in the present case the equations are not linear. Thus, it is not certain that (1) there is a solution; (2) there is not more than one solution; and (3) if a solution exists, that it can be found at reasonable expense.

The development of efficient numerical methods to solve such problems is an area of considerable active research. The various methods and their relative merits are described in an extensive literature. For our needs there are a number of ameliorating circumstances: (1) the dimension of our problem is not great (one to four exponential terms, i.e., between two and eight parameters); (2) we can provide reasonable preliminary estimates for the parameter values by inspection of the data; and (3) we expect only two significant digits in the refined estimates. Moreover, from simple physical intuitive arguments we can assume that a well-conditioned solution to the equations exists (at least for the single exponential fit) if the data resemble an exponential decay. Clearly, when both an initial rise and subsequent fall appear in the data, the procedure should start with two terms and appropriate initial parameter estimates. Thus, we have enough information to prime the program (data plus estimates for parameter values, i.e., D_j and α_j). Alternatively, these estimates may be computed from the data by an initial stage in the program. The program will improve upon the initial parameter estimates (which are only close to the solution of Eq. 3.47) by an iterative procedure that (we hope) will converge to a more nearly exact solution. The principles involved are easily outlined. The function with (unknown) optimal parameter values we seek is

$$\sigma(t) = f(t, \hat{D}_1, D_2, \ldots, D_m, \alpha_1, \alpha_2, \ldots, \alpha_m) \tag{3.48}$$

where D_1, D_2, etc., are the optimal estimates of D_1 and D_2. We have (initially) rough estimates for these values, each of which differs from the optimum by some unknown (positive or negative) amount, say $d_1, d_2, \ldots, d_m$, $\eta_1, \eta_2, \ldots, \eta_m$. By Taylor's formula, we may obtain a series expansion of the optimal solution in terms of our initial estimates $D_1^{(1)}, D_2^{(1)}, \ldots, D_m^{(1)}$, $\alpha_1^{(1)}, \alpha_2^{(1)}, \ldots, \alpha_m^{(1)}$ as follows:

$$\begin{aligned}
f(t, \hat{D}_1, D_2, \ldots, \hat{D}_m, \hat{\alpha}_1, \hat{\alpha}_2, \hat{\alpha}_m) \\
&= f(t, D_1^{(1)} + d_1, D_2^{(1)} + d_2, \ldots, D_m^{(1)} + d_m, \alpha_1^{(1)} + \eta_1, \alpha_m^{(1)} + \eta_2, \ldots, \alpha_m^{(1)} + \eta_m) \\
&= f(t, D_1^{(1)}, D_2^{(1)}, \ldots, D_m^{(1)}, \alpha_1^{(1)}, \alpha_2^{(1)}, \ldots, \alpha_m^{(1)}) \\
&\quad + d_1 \left[\frac{\partial f(t, D_1, D_2, \ldots, D_m, \alpha_1, \alpha_2, \ldots, \alpha_m)}{\partial D_1} \right]^{(1)} \\
&\quad + \cdots + d_m \left[\frac{\partial f(t, D_1, D_2, \ldots, D_m, \alpha_1, \alpha_2, \ldots, \alpha_m)}{\partial D_m} \right]^{(1)} \\
&\quad + \eta_1 \left[\frac{\partial f(t, D_1, D_2, \ldots, D_m, \alpha_1, \alpha_2, \ldots, \alpha_m)}{\partial \alpha_1} \right]^{(1)} \\
&\quad + \cdots + \eta_m \left[\frac{\partial f(t, D_1, D_2, \ldots, D_m, \alpha_1, \alpha_2, \ldots, \alpha_m)}{\partial \alpha_m} \right]^{(1)}
\end{aligned} \tag{3.49}$$

plus higher-order terms involving powers and products of d's and η's and higher-order derivatives. The higher-order terms in the series may be ignored, i.e., they will be small, if the function is smooth in the vicinity of the optimum and the initial estimates are good. With the exclusion of these terms, the problem has become linear in the unknown d's and η's.

For notational convenience we shall denote the set of D's by the symbol D and similarly, the α's by $\underline{\alpha}$ (i.e., $\underline{D}$ denotes $D_1, D_2, \ldots, D_m$ and $\underline{\alpha}$ denotes $\alpha_1, \alpha_2, \ldots, \alpha_m$). Thus, by this notation, $f(t, D_1, D_2, \ldots, D_m, \alpha_1, \alpha_2, \ldots, \alpha_m)$ is simply written $f(t, \underline{D}, \underline{\alpha})$. We now have

$$f(t, \hat{\underline{D}}, \hat{\underline{\alpha}}) \cong f(t, \underline{D}^{(1)}, \underline{\alpha}^{(1)}) + \sum_{j=1}^{m} d_j \frac{\partial}{\partial D_j} f(t, \underline{D}^{(1)}, \underline{\alpha}^{(1)}) + \sum_{j=1}^{m} \eta_j \frac{\partial}{\partial \alpha_j} f(t, \underline{D}^{(1)}, \underline{\alpha}^{(1)}) . \tag{3.50}$$

We seek best values for the d_j and η_j, which will enable us to improve our initial estimates $D_j^{(1)}$ and $\alpha_j^{(1)}$ (i.e., $D_j^{(2)} = D_j^{(1)} + d_j$, $\alpha_j^{(2)} = \alpha_j^{(1)} + \eta_j$). The improved estimates will be used to reinitialize the procedure. The iterations are continued until some convergence criterion is satisfied.

Still unanswered is the matter of solving for the d_i and η_i in each iteration. This may be accomplished by substituting the experimental data itself for the unknown optimum expression in Eq. 3.48 and then solving the resulting linear least-squares problem for the d_j and η_j. That is, at the l-th iteration we seek to minimize $\Sigma_{i=1}^{k} (\varrho_i)^2$ where

$$\varrho_i = \left[f(t_i, \underline{D}^{(l)}, \underline{\alpha}^{(l)}) + \sum_{j=1}^{m} d_j \frac{\partial}{\partial D_j} f(t, \underline{D}^{(l)}, \underline{\alpha}^{(l)}) + \sum_{j=1}^{m} \eta_j \frac{\partial}{\partial \alpha_j} f(t, \underline{D}^{(l)}, \underline{\alpha}^{(l)}) \right] - y_i . \tag{3.51}$$

Repeating the arguments above, now applied to the minimization of $\Sigma_{i=1}^{k} (\varrho_i)^2$, we seek

$$\sum_{i=1}^{k} \varrho_i \frac{\partial \varrho_i}{\partial d_r} = 0 \qquad r = 1, 2, \ldots, m$$

$$\sum_{i=1}^{k} \varrho_i \frac{\partial \varrho_i}{\partial \eta_r} = 0 \qquad r = 1, 2, \ldots, m .$$

This time, however, the problem is linear in the unknown parameters d_r and η_r. The required partial derivatives ($\partial \varrho_i / \partial d_r$, $\partial \varrho_i / \partial \eta_r$) are simply the appropriate derivatives of the function being fitted, evaluated for the present estimates of the D's and α's, as may be seen by inspection of Eq. 3.51. We obtain by substitution

$$\sum_{i=1}^{k} \left\{ \left[f(t_i, \underline{D}, \underline{\alpha}) + \sum_{j=1}^{m} d_j \frac{\partial}{\partial D_j} f(t_i, \underline{D}, \underline{\alpha}) + \sum_{j=1}^{m} \eta_j \frac{\partial}{\partial \alpha_j} f(t_i, \underline{D}, \underline{\alpha}) \right] - y_i \right\} \times \frac{\partial}{\partial D_r} f(t_i, \underline{D}, \underline{\alpha}) = 0 \qquad r = 1, \ldots, m \tag{3.52}$$

$$\sum_{i=1}^{k}\left\{\left[f(t_i, \underline{D}, \underline{\alpha}) + \sum_{j=1}^{m} d_j \frac{\partial}{\partial D_j} f(t_i, \underline{D}, \underline{\alpha}) + \sum_{j=1}^{m} \eta_j \frac{\partial}{\partial \alpha_j} f(t_i, \underline{D}, \underline{\alpha})\right] - y_i\right\}$$

$$\times \frac{\partial}{\partial \alpha_r} f(t_i, \underline{D}, \underline{\alpha}) = 0 \qquad r = 1, \ldots, m. \qquad (3.53)$$

The above equations (3.52 and 3.53) are a system of 2 m linear algebraic equations which, by rearrangement of terms, may be put in the form

$$\sum_{s=1}^{2m} C_s \sum_{i=1}^{k} b_{si}\, a_{ir} = \sum_{i=1}^{k} y_i'\, a_{ir} \qquad r = 1, 2, \ldots, 2\,m \qquad (3.54)$$

where the C's denote the unknown d_i and η_i, the a_{ir}'s are known coefficients (i.e., the partials evaluated at the estimated parameter values) as are the b_{si}; y_i' is the difference between the data point y_i and the fit of the estimate to that point, i.e., $f(t_i, \underline{D}, \underline{\alpha}) - y_i$. In matrix notation, this may be represented as

$$P\,C = Z \qquad \text{Solution: } C = P^{-1} Z\,.$$

The system of linear algebraic equations is easily solved by manipulation requiring only elementary arithmetic and logical operations provided that P is not ill conditioned (singular or nearly so) (Note 3.3). It is frequently the case that the individual datum in a series are not known with equal precision. For example, the measurement technique may be characterized by constant relative error rather than by constant absolute error. Under such circumstances, the regression should be applied to minimize the sum

$$\Phi = \sum_{i=1}^{k} w_i\, \varrho_i^2$$

in which the importance of fitting each datum is graded by an *a priori* estimate of its accuracy. The appropriate weight is proportional to the reciprocal of the estimated variance of the observational technique at that point. Note that only an estimate of the relative weight of each point is required. Alternatively, in the absence of variance estimates on a point by point basis, the weights should be uniform (i.e., $w_i \cong 1$) for an experiment characterized by hypothetical constant absolute error, and in the proportion $w_i \cong (1/y_i)^2$, where constant relative error is a preferable assumption. The choice of the w_i can significantly affect the resulting "optimal" parameter estimates (see Note 5.1) and thus an effort should be made to determine the procedure appropriate to each application.

Having obtained "best" values for an m exponential approximation, the program may then compare the residual deviation of the data about the analytic curve with a user-supplied estimate of the variance resulting from measurement error and other sources (if this is available).

The basis for comparison might be the F test. The objective is, of course, to determine whether the m terms are a satisfactory representation of the trend in the data (i.e., changes beyond those resulting from methodologic and other random causes). It may often be the case that no estimate for these latter factors is available. Under such circumstances, the program could be used to fit m + 1 exponential terms, i.e., two additional parameters to the data (using either some adjustment of the previous D_j and α_j or the peel-off method for initial approximations). Having completed this exercise, the remaining deviation of the data (which will, in general, decrease) may then be compared with the previous result (by, for example, the F test) to determine if the improvement in the fit justifies the hypothesis of an additional term. If the additional term significantly improved matters, then yet another term must be similarly investigated and so forth until the addition of a term fails the test, or a reasonable *a priori* limit (e.g., three or four terms) is reached. The number of terms will, of course, depend on the data (i.e., the number of observations, their accuracy and the nature of the underlying system). These matters are illustrated in Note 5.1 in Chapter 5.

In addition to calculating the estimates of "least-square" parameter values, the procedure outlined above can be adapted to provide some indication of the "confidence" that may be placed in the parameters and parameter values. The most easily computed estimate of this nature is the parameter asymptotic variance and covariance estimates obtained from the P^{-1} matrix. At the last stage of the iterated procedure, the element of this matrix represents the square of the partial derivatives of the regression parameters with respect to the computed optimal approximation of the data.

An easily derived relation in probability theory is that if

$$X = c\,Y$$

then

$$\mathrm{Var}(X) = c^2\,\mathrm{Var}(Y)$$

where X, Y are random variables and c a constant coefficient.

More generally, when X is some function of Y

$$X = f(Y)$$

$$\mathrm{Var}(X) \cong \left(\frac{\delta f}{\delta Y}\right)^2 \mathrm{Var}(Y)$$

or, for the multivariate case, where Y_i are independently distributed

$$X = f(Y_1, Y_2, \ldots, Y_k)$$

$$\mathrm{Var}(X) \cong \sum_{i=1}^{k} \left(\frac{\delta f}{\delta Y_i}\right)^2 \mathrm{Var}(Y_i)\,.$$

The elements of the P^{-1} matrix thereby permit estimation of the error in the fitted parameters.

In summary, a standard procedure has been outlined in this section by which the problem of estimating least-squares best-fit values for parameters which appear nonlinearly in an analytic function (e.g., $D\,e^{-\alpha t}$), may be converted to an iterated linear least-squares problem for the gradual refinement of initial parameter estimates. The linear problem (which must be solved repeatedly until calculated incremental corrections converge) is, in general, easily solved by arithmetic methods. As has been alluded to repeatedly, an important factor in the performance of the method is the precision of the initial estimate. This may make the difference between success and failure and, at the very least, will significantly influence the machine time required to find the optimum. For these reasons, the user is well advised to supply good starting values. This might be accomplished by use of the peel-off method (performed either by hand or by a computer-implemented program). The increasing availability and economy of "time-sharing" remote terminals and interactive programs removes the last obstacle to routine use of these methods.

Note 3.3

The procedure is simply that which is commonly used for the solution of systems of simultaneous linear algebraic equations.

$$\begin{aligned} &w_{11} c_1 + w_{12} c_2 + w_{13} c_3 + \cdots + w_{1,2m} c_{2m} = Z_1 \\ &w_{21} c_1 + w_{22} c_2 + w_{23} c_3 + \cdots + w_{2,2m} c_{2m} = Z_2 \\ &\cdots\cdots\cdots\cdots\cdots\cdots \\ &w_{2m,1} c_1 + w_{2m,2} c_2 + w_{2m,3} c_3 + \cdots + w_{2m,2m} c_{2m} = Z_{2m} . \end{aligned} \tag{3.55}$$

The objective is to obtain, by appropriate elementary operations, an explicit expression for $c_1, c_2, c_3, \ldots, c_{2m}$ in terms of the known coefficients w_{rs} and the Z_r. Provided that the equations are ordered so that $w_{11} \neq 0$, division of both sides of the first equation by w_{11} yields:

$$c_1 + \frac{w_{12}}{w_{11}} c_2 + \frac{w_{13}}{w_{11}} c_3 + \cdots + \frac{w_{1,2m}}{w_{11}} c_{2m} = \frac{Z_1}{w_{11}} . \tag{3.56}$$

Note that neither the division nor the operations described below affect the (as yet unknown) values of the c_i, which satisfy the 2 m expressions. Thus, any solution of the transformed system is a solution of the original system. The equality (Eq. 3.56) multiplied by the coefficient of the c_1 term in each subsequent equation, may then, in turn, be subtracted from the corresponding equation to produce

$$c_1 + \frac{w_{12}}{w_{11}} c_2 + \frac{w_{13}}{w_{11}} c_3 + \cdots + \frac{w_{1,2m}}{w_{11}} c_{2m} = \frac{Z_1}{w_{11}} = Z_1^{(1)}$$

$$\left(w_{22} - w_{21} \frac{w_{21}}{w_{11}}\right) c_2 + \left(w_{23} - w_{21} \frac{w_{13}}{w_{11}}\right) c_3 + \cdots + \left(w_{2,2m} - w_{21} \frac{w_{1,2m}}{w_{11}}\right) c_{2m}$$
$$= Z_2 - w_{21} \frac{Z_1}{w_{11}} = Z_2^{(1)}$$

$$\cdots \qquad \cdots \qquad \cdots$$

$$\left(w_{2m\,2} - w_{2m\,1} \frac{w_{12}}{w_{11}}\right) c_2 + \left(w_{2m\,3} - w_{2m\,1} \frac{w_{13}}{w_{11}}\right) c_3 + \cdots$$
$$+ \left(w_{2m\,2m} - w_{2m\,1} \frac{w_{2m\,2m}}{w_{11}}\right) c_{2m} = Z_{2m} - w_{2m\,1} \frac{Z_1}{w_{11}} = Z_{2m}^{(1)}$$

thereby eliminating the variable c_1 from every equation except one (i.e., all but one of the new coefficients $w_{i1}^{(1)}$ obtained in this way are zero). Clearly, the same procedure can then be applied to the transformed equations to obtain a system in which c_2 is eliminated from every equation below the second and so forth until the 2 m equations are of the form

$$c_1 + w_{12}^{(1)} c_2 + w_{13}^{(1)} c_3 \ldots + w_{1,2m}^{(1)} c_{2m} = Z_1^{(1)}$$

$$c_2 + w_{23}^{(2)} c_3 \ldots + w_{2,2m}^{(2)} c_{2m} = Z_2^{(2)}$$

$$\ldots \qquad \ldots$$

$$c_{2m-1} + w_{2m-1,2m}^{(2m-1)} c_{2m} = Z_{2m-1}^{(2m-1)}$$

$$c_{2m} = Z_{2m}^{(2m)}$$

where $Z^{(a)}$ denotes the expression resulting from the sequence of arithmetic operations performed on the initial Z. The explicit solution for c_{2m} may now be substituted in the expression immediately above it, to obtain c_{2m-1} and so forth. The method illustrates the fundamental simplicity of the so-called elimination techniques; however, it is important to note that refinements employed in the popular versions of such algorithms offer a significant reduction in computation expense. By exploiting special properties of a particular problem (e.g., real symmetric and sparse coefficient matrices) further economies may be achieved. Moreover, for a given arithmetic precision, improvements in solution accuracy may be obtained by careful ordering of the equations at each stage.

Chapter 4

Tracer Kinetics in Compartmental Models

This chapter presents some aspects of the classic approach to the interpretation of tracer experiments. A well-defined system of pools at the steady state is postulated and the number of pools in the system is assumed to be known or experimentally determinable. The mixing in each of these pools is considered to be instantaneous and the rate of exit of labeled molecules to be proportional to their concentrations in the pool.

On the basis of these assumptions, the specific activity functions in each pool following the instantaneous introduction of a tracer into one of them can be calculated in terms of parameters describing the movement of material into, out of, and between the pools. Conversely, and more importantly, these parameters may be calculated from tracer data.

Following the pattern used in the preceding chapters, systems of 1, 2, 3, and n pools at the steady state are considered. The symbol r_{ij} is used in this chapter to denote the rate of direct transfer from pool i to pool j. Rates are represented by r's instead of the v's used in the preceding chapters in order to emphasize the postulate that there are no intermediates involved in the process described by a rate r_{ij} and that the total number of pools in the system is known.

The present analysis is purely theoretical and the number of ideal pools in the system is given as known data. The applicability of this analysis to interpret tracer data obtained from biologic systems is discussed in Chapter 5.

Rate Constants (k's)

If the rate of transfer of material from pool i to pool j is denoted by r_{ij} (µmole/hr), the rate constant (k_{ij}) corresponding to this process is defined as $k_{ij} = r_{ij}/M_i$, where M_i is the amount of the compound in pool i. According to this definition, a value of k_{ij} of 0.01 h^{-1} indicates that the transfer from pool i to pool j occurs, at the steady state, at a rate equivalent to the transfer of 1/100 of the content of pool i to pool j per hour.

Since labeled and unlabeled molecules in a pool are assumed to behave identically (a basic postulate in tracer experiments), the rate of transfer of isotope from pool i to pool j is given by the product $k_{ij}\,y_i$, if y_i is the total amount of isotope in pool i. At the steady state, k_{ij} is constant since r_{ij} and M_i are constant. These definitions imply postulating that the unidirectional transfer of isotope follows first-order kinetics, i.e., that the rate of transfer of the isotope is proportional to the concentration of the isotope in the pool (see Note 3.2).

The general expressions describing the rate of change of the amount of an isotope in pool i is given by the difference between the rates at which the isotope enters and leaves the pool. Thus, if y_i^m is the amount of an isotope introduced into pool m that is present in pool i, it follows that dy_i^m/dt equals the difference between the rate of entry of isotope m into pool i and the rate of exit of the isotope from the pool. If the rate of exit is larger than the rate of entry, then dy_i^m/dt is negative and corresponds to a decrease in the isotope content of the pool. The rate of entry of isotope m into pool i equals the sum of all the rates of transfer to pool i from pools directly communicating with it. Each of these rates can be expressed as $k_{ji}\, y_j^m$, where j is any of the pools in the system except pool i. Of course, if a pool j does not communicate directly with pool i, then $k_{ji} = 0$ even if pool j supplies isotope to pool i through the intermediacy of other pools. The rate of exit of the isotope m from pool i is given by the product $(-k_{ii})\, y_i^m$ where $-k_{ii} = (-r_{ii})/M_i$ is the rate constant corresponding to the rate of removal of the compound from pool i by all processes. The minus sign is attached to k_{ii} and r_{ii} for convenience in writing the equations, as is evident from Eq. 4.1. It should be noted that $-k_{ii}$ and $-r_{ii}$ are positive numbers.

It then follows that the rate of change of the isotope content of pool i can be described by the following equation:

$$\frac{dy_i^m}{dt} = \sum_{\substack{j=1 \\ j \neq i}}^{n} k_{ji}\, y_j^m - (-k_{ii})\, y_i^m = \sum_{j=1}^{n} k_{ji}\, y_j^m\,. \tag{4.1}$$

Fractional Rates (h's)

A difficulty with Eq. 4.1 is that it involves values corresponding to the total amount of isotope in a pool, a quantity not easily measurable. It may be more convenient to establish equations in terms of specific activities, which can be measured in any sample of the pool, even when the size of the pool is not known. To achieve the conversion of Eq. 4.1 to an equation in specific activities, a constant h, defined by the relation $h_{ij} = r_{ij}/M_j$ or

$$h_{ij} = \frac{M_i}{M_j}\, k_{ij} \qquad i, j = 1, 2, \ldots, n \tag{4.2}$$

can be used (MANN and GURPIDE, 1969 a; HEARON, 1963).

Since the specific activity of pool i with respect to an isotope introduced into pool m (σ_i^m) equals the ratio y_i^m/M_i, Eq. 4.1 is equivalent to

$$M_i \frac{d\sigma_i^m}{dt} = \sum_{j=1}^{n} k_{ji}\, M_j\, \sigma_j^m$$

or

$$\frac{d\sigma_i^m}{dt} = \sum_{j=1}^{n} \frac{k_{ji} M_j}{M_i} \sigma_j^m$$

and, according to Eq. 4.2,

$$\frac{d\sigma_i^m}{dt} = \sum_{j=1}^{n} h_{ji} \sigma_j^m \qquad i = 1, 2, \ldots, n. \tag{4.3}$$

The similarity of Eqs. 4.1 and 4.3 may be noted.

A. One-Pool System

This case is illustrated in Fig. 4.1, which differs from Fig. 1.1 by the absence of pathways of recycle. Note that in a one-pool system $-r_{11} = r_{10}$ and $r_{01} = PR_1$.

Fig. 4.1. Representation of a single pool system

Rapid Injection of the Tracer

The differential equation describing the rate of change of the specific activity in the pool following the instantaneous administrations of S* dpm of tracer into the pool is, according to Eq. 4.3,

$$\frac{d\sigma_1^1}{dt} = h_{11} \sigma_1^1 \tag{4.4}$$

or, since $h_{11} = k_{11}$ by Eq. 4.2,

$$\frac{d\sigma_1^1}{dt} = k_{11} \sigma_1^1 = -k_{10} \sigma_1^1. \tag{4.5}$$

Equation 4.5 can be obtained directly by noting that the rate at which isotope leaves the pool is $k_{10} y_1^1$. Therefore, the rate of change in the isotope content in the pool is given by the expression

$$\frac{dy_1^1}{dt} = -k_{10} y_1^1. \tag{4.6}$$

Division of this equation by M_1 yields Eq. 4.5.

Equation 4.5 can be written as $d\sigma_1^1/\sigma_1^1 = -k_{10}\,dt$ or, since $d\sigma_1^1/\sigma_1^1 = d\ln\sigma_1$, it is also equivalent to the following:

$$d\ln\sigma_1^1 = -k_{10}\,dt\,.$$

Integration of this equation yields the result $\ln\sigma_1^1 = -k_{10}\,t + \text{constant}$. Since at $t=0$, $\sigma_1^1 = (\sigma_1^1)_0$ the constant of integration must equal the natural logarithm of the specific activity at time zero, i.e.,

$$\ln\sigma_1^1 = -k_{10}\,t + \ln(\sigma_1^1)_0 \tag{4.7}$$

or

$$\ln\frac{\sigma_1^1}{(\sigma_1^1)_0} = -k_{10}\,t \tag{4.8}$$

and, using the definition of natural logarithms, $\sigma_1^1/(\sigma_1^1)_0 = e^{-k_{10}t}$. Then,

$$\sigma_1^1 = (\sigma_1^1)_0\,e^{-k_{10}t}\,. \tag{4.9}$$

"Half-Life" ($t_{1/2}$)

Equation 4.7 indicates that a plot of the logarithm of the specific activity versus time, e.g., a semilogarithmic plot of $\sigma_1^1 = f(t)$, yields a straight line, with a slope equal to $-k_{10}$ and an intercept equal to $(\sigma_1^1)_0$, as shown in Fig. 4.2. The slope of the line is usually expressed in terms of a "half-life" ($t_{1/2}$), i.e., the time necessary to reduce a specific activity to half its value.

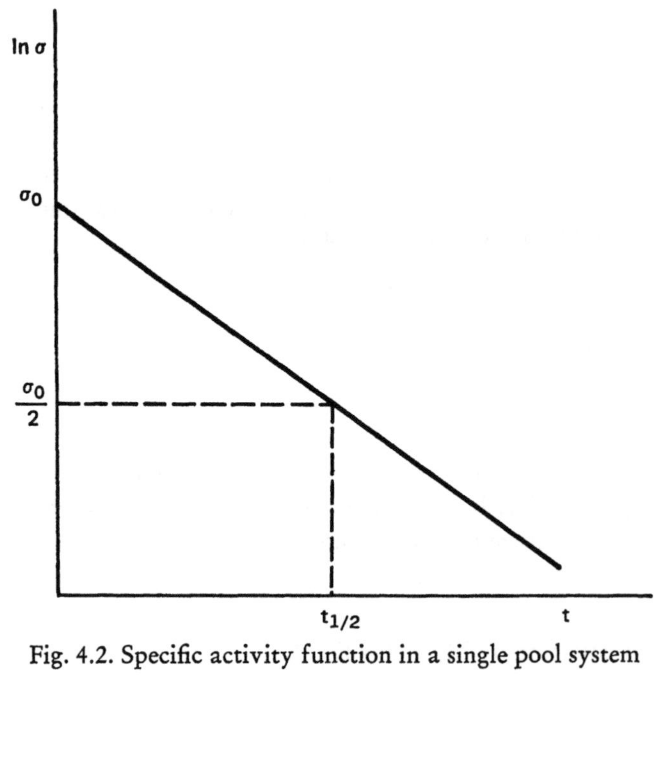

Fig. 4.2. Specific activity function in a single pool system

The relation between $t_{1/2}$ and k_{10} is clear from Eq. 4.8. Thus, since $\sigma_1^1 = (\sigma_1^1)_0/2$ when $t = t_{1/2}$, then $\ln \frac{1}{2} = -k_{10}\, t_{1/2}$ or

$$k_{10} = \frac{\ln 2}{t_{1/2}} = 0.693/t_{1/2}\,. \tag{4.10}$$

The half-life can be easily read from the graph as the time period between any two values of specific activities differing by a factor of two.

Time Constant (τ)

If, at the steady state, an amount equivalent to a fraction k_{10} of M_1 is irreversibly removed per unit time, an amount equal to M_1 will be removed in an interval τ. Then, $\tau/M_1 = 1/(k_{10}\,M_1)$ or $\tau = 1/k_{10}$. This interval is called the time constant of the system. It is apparent from Eq. 4.9 that

$$\sigma_1^1 = (\sigma_1^1)_0\, e^{-(t/\tau)}$$

and that the value of the specific activity will fall by a factor of $1/e$ over the interval τ.

Infusion of the Tracer at a Constant Rate

The relationship shown in Chapter 3 (Eq. 3.8) serves to calculate the specific activity of a compound in a pool at any time during infusion of the tracer from data obtained after rapid injection of the tracer. Thus, if $a_{1\tau}^1$ denotes the specific activity of the compound in the pool at time τ after starting the infusion of the tracer at a rate I_1^*, it follows from Eqs. 3.8 and 4.9 that

$$a_{1\tau}^1 = \frac{I_1^*}{S_1^*}\int_0^\tau \sigma_1^1\; dt = \frac{I_1^*}{S_1^*}(\sigma_1^1)_0 \int_0^\tau e^{-k_{10}t}\, dt$$

or, since $(\sigma_1^1)_0 = S_1^*/M_1$,

$$a_{1\tau}^1 = \frac{I_1^*}{M_1}\left[\frac{e^{-k_{10}\tau}}{-k_{10}} - \frac{1}{-k_{10}}\right] = \frac{I_1^*}{v_{10}}\left(1 - e^{-k_{10}\tau}\right). \tag{4.11}$$

When $t \to \infty$, the specific activity at the steady state ($a_{1\infty}^1$, or simply a_1^1) is $a_1^1 = I_1^*/v_{10}$, as already indicated (Eq. 2.1). It then follows from Eq. 4.11 that

$$a_{1\infty}^1 - a_{1t}^1 = \frac{I_1^*}{k_{10}\,M_1}\, e^{-k_{10}t}$$

or

$$\ln\left(a_{1\infty}^1 - a_{1t}^1\right) = -k_{10}\,t + \ln \frac{I_1^*}{k_{10}\,M_1}\,.$$

It is apparent that a semilogarithmic plot of the differences of the values of specific activities at the steady state and at various times of infusion results in a straight line from which k_{10} and M_1 can be estimated.

B. Two-Pool Systems

This case is illustrated by the model in Fig. 4.3.

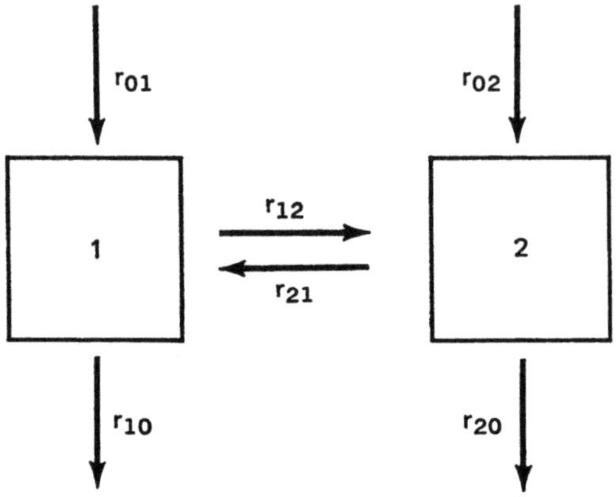

Fig. 4.3. Representation of a two-pool system

Rapid Injection of the Tracer

The differential equations describing the rates of change of the specific activities in pools 1 and 2 following the instantaneous injection of a tracer into pool 1 are the following (see Eq. 4.3):

$$\frac{d\sigma_1^1}{dt} = h_{11}\,\sigma_1^1 + h_{21}\,\sigma_2^1 \tag{4.12}$$

$$\frac{d\sigma_2^1}{dt} = h_{12}\,\sigma_1^1 + h_{22}\,\sigma_2^1\,. \tag{4.13}$$

Equations 4.12 and 4.13 constitute a set of two linear differential equations with constant coefficients. The solution for this simple system, e.g., by using Laplace transforms, can be found in textbooks on differential equations (Agnew, 1960; Rainville and Bedient, 1969), i.e.,

$$\sigma_1^1 = D_{11}\,e^{-\alpha_1 t} + D_{12}\,e^{-\alpha_2 t} \tag{4.14}$$

$$\sigma_2^1 = D_{21}\,e^{-\alpha_1 t} + D_{22}\,e^{-\alpha_2 t}\,. \tag{4.15}$$

Minus signs are attached to the exponents (in Eqs. 4.14 and 4.15). Therefore, neither α_1 nor α_2 in Eq. 4.14 may be zero or negative, for in the former case the specific activity would not approach zero, and in the latter case σ_1^1 would become infinitely large.

The reader may easily verify that these are the solutions of the system of Eqs. 4.12 and 4.13. Substituting $d\sigma_1^1/dt$, σ_1^1, and σ_2^1 in Eq. 4.12 by the expressions derived from Eqs. 4.14 and 4.15, it follows that

$$\begin{aligned} -\alpha_1 D_{11}\,e^{-\alpha_1 t} - \alpha_2 D_{12}\,e^{-\alpha_2 t} &= h_{11}(D_{11}\,e^{-\alpha_1 t} + D_{12}\,e^{-\alpha_2 t}) + h_{21}(D_{21}\,e^{-\alpha_1 t} + D_{22}\,e^{-\alpha_2 t}) \\ &= (h_{11} D_{11} + h_{21} D_{21})\,e^{-\alpha_1 t} + (h_{11} D_{12} + h_{21} D_{22})\,e^{-\alpha_2 t}\,. \end{aligned}$$

This identity requires that

$$-\alpha_1 D_{11} = h_{11} D_{11} + h_{21} D_{21} \tag{4.16}$$

and

$$-\alpha_2 D_{12} = h_{11} D_{12} + h_{21} D_{22} . \tag{4.17}$$

Similarly, it follows from Eqs. 4.13, 4.14, and 4.15 that

$$-\alpha_1 D_{21} = h_{12} D_{11} + h_{22} D_{21} \tag{4.18}$$

and

$$-\alpha_2 D_{22} = h_{12} D_{12} + h_{22} D_{22} . \tag{4.19}$$

It is evident that there are values of D_{11}, D_{12}, D_{21}, and D_{22}, which would satisfy Eqs. 4.16 to 4.19 and that, therefore, 4.14 and 4.15 are the solutions of the systems of Eqs. 4.12 and 4.13.

Equations 4.16 and 4.18 can be written as follows:

$$(h_{11} + \alpha_1) D_{11} + h_{21} D_{21} = 0 \tag{4.20}$$

and

$$h_{12} D_{11} + (h_{22} + \alpha_1) D_{21} = 0 . \tag{4.21}$$

Similarly, Eqs. 4.17 and 4.19 are, respectively,

$$(h_{11} + \alpha_2) D_{12} + h_{21} D_{22} = 0 \tag{4.22}$$

and

$$h_{12} D_{12} + (h_{22} + \alpha_2) D_{22} = 0 . \tag{4.23}$$

In order to have values of D_{11} and D_{12}, and of D_{21} and D_{22}, which are different from 0, it is necessary that the determinant of the coefficients in these two systems of equations be zero, i.e.,

$$\begin{vmatrix} h_{11} + \alpha_1 & h_{21} \\ h_{12} & h_{22} + \alpha_1 \end{vmatrix} = 0 \quad \text{and} \quad \begin{vmatrix} h_{11} + \alpha_2 & h_{21} \\ h_{12} & h_{22} + \alpha_2 \end{vmatrix} = 0 .$$

Therefore, the coefficients of the exponents in Eqs. 4.14 and 4.15, (α_1 and α_2), are the roots of the quadratic equation in α given by the following determinant:

$$\begin{vmatrix} h_{11} + \alpha & h_{21} \\ h_{12} & h_{22} + \alpha \end{vmatrix} = 0$$

i.e., $\alpha^2 + (h_{11} + h_{22}) \alpha + (h_{11} h_{22} - h_{12} h_{21}) = 0$.

A quadratic equation $x^2 + p x + q = 0$ with roots x_1 and x_2 may be written in factorial form as $(x - x_1)(x - x_2) = 0$. Multiplication yields $x^2 - (x_1 + x_2) x + x_1 x_2$. Thus, it is clear that in an equation of this type the sum of the roots equals the coefficient of the variable and the product of the roots equals the independent term. Then,

$$\alpha_1 + \alpha_2 = -(h_{11} + h_{22}) = -(k_{11} + k_{22}) \tag{4.24}$$

$$\alpha_1 \alpha_2 = h_{11} h_{22} - h_{12} h_{21} = k_{11} k_{22} - k_{12} k_{21} . \tag{4.25}$$

If α_1 is chosen to be larger than α_2, it follows that

$$\alpha_1 = \frac{-(h_{11} + h_{22}) + \sqrt{(h_{11} - h_{22})^2 + 4 h_{12} h_{21}}}{2} \tag{4.26}$$

and

$$\alpha_2 = \frac{-(h_{11}+h_{22}) - \sqrt{(h_{11}-h_{22})^2 + 4\, h_{12}\, h_{21}}}{2}. \tag{4.27}$$

Other forms of Eqs. 4.26 and 4.27 are the following:

$$\alpha_1 = \frac{-(k_{11}+k_{22}) + \sqrt{(k_{11}-k_{22})^2 + 4\, k_{12}\, k_{21}}}{2} \tag{4.28}$$

and

$$\alpha_2 = \frac{-(k_{11}+k_{22}) - \sqrt{(k_{11}-k_{22})^2 + 4\, k_{12}\, k_{21}}}{2}. \tag{4.29}$$

If either k_{12} or k_{21} equal zero (as is the case when the transfer of material between pools 1 and 2 is irreversible) then

$$\alpha_1 = -k_{11}$$

and

$$\alpha_2 = -k_{22}.$$

If $\alpha_1 = \alpha_2$, Eqs. 4.14 and 4.15 are not the solutions for the system of Eqs. 4.12 and 4.13. This particular case is discussed in another section in this chapter (Eqs. 4.48 and 4.49).

Equations 4.24 to 4.27 give the values of α_1 and α_2 in terms of h's or k's. Therefore, the α's are fixed by the system. However, the coefficients D_{11}, D_{12}, D_{21}, and D_{22} depend upon the initial conditions, namely, the amount of isotope in pools 1 and 2 at time zero.

Letting $(\sigma_1^1)_0$ denote the value of σ_1^1 at $t=0$ and $(d\sigma_1^1/dt)_0$ denote the value of $d\sigma_1^1/dt$ at $t=0$ [with similar interpretations for $(\sigma_2^1)_0$ and $(d\sigma_2^1/dt)_0$] it follows from Eq. 4.14 that

$$D_{11} + D_{12} = (\sigma_1^1)_0 \tag{4.30}$$

and, from Eq. 4.12,

$$\left(\frac{d\sigma_1^1}{dt}\right)_0 = h_{11}(\sigma_1^1)_0 + h_{21}(\sigma_2^1)_0 = h_{11}(\sigma_1^1)_0 \tag{4.31}$$

if, as is assumed here, $(\sigma_2^1)_0 = 0$, i.e., the pools are unlabeled before the injection of the tracer. Setting $t=0$ in the derivative of Eq. 4.14, it is found that

$$(d\sigma_1^1/dt)_0 = -\alpha_1 D_{11} - \alpha_2 D_{12}$$

or by Eq. 4.31,

$$-\alpha_1 D_{11} - \alpha_2 D_{12} = h_{11}(\sigma_1^1)_0. \tag{4.32}$$

The solution of the simultaneous Eqs. 4.30 and 4.32 by determinants using Cramer's rule, for instance, is

$$D_{11} = -(\sigma_1^1)_0 \frac{\alpha_2 + h_{11}}{\alpha_1 - \alpha_2} \tag{4.33}$$

and

$$D_{12} = (\sigma_1^1)_0 \frac{\alpha_1 + h_{11}}{\alpha_1 - \alpha_2}. \tag{4.34}$$

Similarly, it follows from Eq. 4.13 that

$$\left(\frac{d\sigma_2^1}{dt}\right)_0 = h_{12}(\sigma_1^1)_0 + h_{22}(\sigma_2^1)_0 = h_{12}(\sigma_1^1)_0. \tag{4.35}$$

Setting $t=0$ in Eqs. 4.14 and 4.15 and in the derivative of Eq. 4.15 one obtains

$$D_{21}+D_{22}=(\sigma_2^1)_0=0 \tag{4.36}$$

and

$$-\alpha_1 D_{21}-\alpha_2 D_{22}=\left(\frac{d\sigma_2^1}{dt}\right)_0=h_{12}(\sigma_1^1)_0 . \tag{4.37}$$

Therefore,

$$D_{21}=-(\sigma_1^1)_0 \frac{h_{12}}{\alpha_1-\alpha_2} \tag{4.38}$$

and

$$D_{22}=(\sigma_1^1)_0 \frac{h_{12}}{\alpha_1-\alpha_2} . \tag{4.39}$$

Thus, Eqs. 4.14 and 4.15 may be written as

$$\sigma_1^1=(\sigma_1^1)_0 \frac{-(\alpha_2+h_{11})e^{-\alpha_1 t}+(\alpha_1+h_{11})e^{-\alpha_2 t}}{\alpha_1-\alpha_2} \tag{4.40}$$

$$\sigma_2^1=(\sigma_1^1)_0 \frac{h_{12}(-e^{-\alpha_1 t}+e^{-\alpha_2 t})}{\alpha_1-\alpha_2} . \tag{4.41}$$

Determination of α's and D's from Experimental Data

Recall that Eq. 4.14 states that

$$\sigma_1^1=D_{11}e^{-\alpha_1 t}+D_{12}e^{-\alpha_2 t}$$

where α_1 was chosen to be larger than α_2.

The parameters α_1, α_2, D_{11}, and D_{12} can be determined from experimental data, e.g. using the graphic method described in Chapter 3.

The values of D_{21} and D_{22} (Eq. 4.15) can be similarly obtained from the experimental values of σ_2^1. D_{22} is determined from the tail portion of the curve of σ_2^1, when the term $D_{21}e^{-\alpha_1 t}$ is insignificant with respect to $D_{22}e^{-\alpha_2 t}$ and a semilogarithmic plot of σ_2^1 against t results in a straight line. It is known already from Eq. 4.36 that $D_{21}=-D_{22}$. Consequently, it is not necessary to analyze σ_2^1 for early values of t.

It should be noted that most of the information about the system is derived from data obtained from pool 1. As can be seen from Eq. 4.15, for large values of t, $\sigma_2^1 \sim D_{22}e^{-\alpha_2 t}$. Since α_2 is determinable from data obtained from pool 1, one late value of σ_2^1 suffices to determine D_{22}.

Analysis of Two-Pool Systems with Equal Exponential Constants ($\alpha_1 = \alpha_2$)

The evaluation of the h_{ij} described above is based on the differential equations 4.12 and 4.13 and also on the solutions to these differential equations given by Eqs. 4.14 and 4.15. If, however, $\alpha_1 = \alpha_2$, these are not the correct solutions.

Before giving the solutions when $\alpha_1 = \alpha_2$, it is worthwhile to consider what properties the h_{ij} must have to cause α_1 to equal α_2.

It follows from Eqs. 4.26 and 4.27 that $\alpha_1 = \alpha_2$, if, and only if,

$$(h_{11} - h_{22})^2 + 4\, h_{12}\, h_{21} = 0\,. \tag{4.42}$$

Since each of the terms on the left side of Eq. 4.42 is positive, each must vanish if the sum of the two is zero. Thus, h_{11} must equal h_{22} and either h_{12} or h_{21} must equal zero. If the tracer is injected into pool 1, only the case in which $h_{21} = 0$ need be considered since otherwise pool 2 would not be labeled.

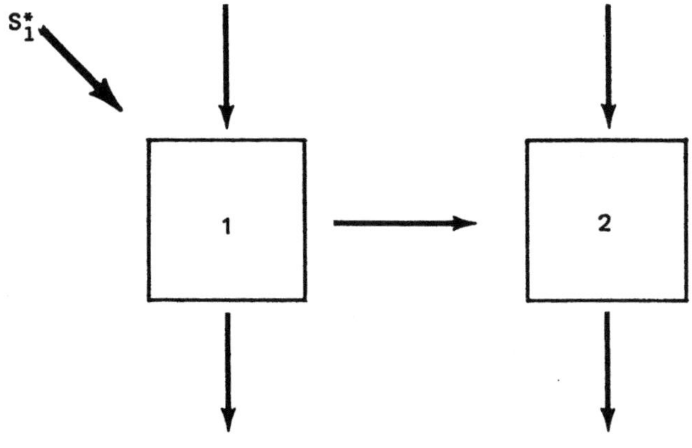

Fig. 4.4. Two pool system in which $r_{21} = 0$ and tracer enters through pool 1 (see case 6)

When $h_{21} = 0$ (Fig. 4.4), Eqs. 4.12 and 4.13 may be written as

$$\frac{d\sigma_1^1}{dt} = h_{11}\, \sigma_1^1 \tag{4.43}$$

$$\frac{d\sigma_2^1}{dt} = h_{12}\, \sigma_1^1 + h_{11}\, \sigma_2^1\,. \tag{4.44}$$

The general form of the solution of these differential equations is (Rainville and Bedient, 1969)

$$\sigma_1^1 = E_{11}\, e^{-at}\,. \tag{4.45}$$

$$\sigma_2^1 = E_{21}\, e^{-at} + E_{22}\, t\, e^{-at}\,. \tag{4.46}$$

Since at $t=0$, $\sigma_2^1=0$, it follows that $E_{21}=0$ and therefore $\sigma_2^1=E_{22}\,te^{-\alpha t}$. Differentiation of this equation yields

$$E_{22}=\left(\frac{d\sigma_2^1}{dt}\right)_0=h_{12}\,(\sigma_1^1)_0\,. \tag{4.47}$$

Consequently, Eqs. 4.45 and 4.46 can be written as follows:

$$\sigma_1^1=(\sigma_1^1)_0\,e^{-\alpha t} \tag{4.48}$$

$$\sigma_2^1=(\sigma_1^1)_0\,h_{12}\,t\,e^{-\alpha t}\,. \tag{4.49}$$

The values of α and $(\sigma_1^1)_0$ can be obtained from semilogarithmic plots of σ_1^1 as described before. It follows from Eq. 4.49 that $\ln\sigma_2^1=-\alpha t+\ln t+\ln(\sigma_1^1)_0+\ln h_{12}$. This equation indicates that a straight line is not to be expected if $\ln\sigma_2^1$ is plotted against t. However, if $\ln(\sigma_2^1/t)$ is plotted against t a straight line results, since

$$\ln\frac{\sigma_2^1}{t}=-\alpha t+\ln(\sigma_1^1)_0+\ln h_{12}\,. \tag{4.50}$$

Dividing Eq. 4.48 by Eq. 4.49, it follows that

$$\frac{\sigma_1^1}{\sigma_2^1}=\frac{1}{h_{12}\,t}\,. \tag{4.51}$$

Therefore, h_{12} can be calculated from any pair of simultaneous values of σ_1^1 and σ_2^1. Finally, it follows from Eqs. 4.43 and 4.48 that

$$h_{11}=h_{22}=-\alpha\,. \tag{4.52}$$

Analysis of a Two-Pool Closed System

A closed system is characterized by the conditions $r_{10}=r_{20}=r_{01}=r_{02}=0$, i.e. no material enters of leaves the system.

In this case, $-k_{11}=k_{12}$ and $-k_{22}=k_{21}$. Therefore, from Eq. 4.25, $\alpha_1\alpha_2=0$ or $\alpha_2=0$, and from Eq. 4.24,

$$\alpha=k_{12}+k_{21}\,.$$

Furthermore, since $k_{12}M_1=k_{21}M_2$, it follows that $k_{12}=(M_2/M_1)\,k_{21}=h_{21}$ and $k_{21}=(M_1/M_2)\,k_{12}=h_{12}$.

Consequently, from Eqs. 4.40 and 4.41,

$$\sigma_1^1=(\sigma_1^1)_0\,\frac{k_{12}\,e^{-\alpha t}+k_{21}}{k_{12}+k_{21}}=(\sigma_1^1)_0\,\frac{k_{21}}{k_{12}+k_{21}}\left(1+\frac{k_{12}}{k_{21}}\,e^{-\alpha t}\right) \tag{4.53}$$

and

$$\sigma_2^1=(\sigma_1^1)_0\,\frac{k_{21}}{k_{12}+k_{21}}\,(1-e^{-\alpha t})\,. \tag{4.54}$$

When $t \to \infty$, then $e^{-\alpha t} \to 0$ and the specific activities in both pools become identical and equal to the following expression:

$$(\sigma_1^1)_\infty = (\sigma_2^1)_\infty = (\sigma_1^1)_0 \frac{k_{21}}{k_{12}+k_{21}} . \tag{4.55}$$

Equation 4.53 can then be written in the form

$$\sigma_1^1 - (\sigma_1^1)_\infty = (\sigma_1^1)_\infty \frac{k_{12}}{k_{21}} e^{-\alpha t} .$$

Therefore,

$$\ln \frac{\sigma_1^1 - (\sigma_1^1)_\infty}{(\sigma_1^1)_\infty} = -\alpha t + \ln \frac{k_{12}}{k_{21}} .$$

Consequently, a semilogarithmic plot of the differences between the specific activity in pool 1 and the specific activity at equilibrium relative to the equilibrium value yields a straight line with a slope α and an intercept k_{12}/k_{21}. Since $\alpha = k_{12} + k_{21}$, the rate constants k_{12} and k_{21} can be determined from data obtained from pool 1.

The pool sizes M_1 and M_2 can also be estimated from the same data since, from Eq. 4.55,

$$(\sigma_1^1)_0 = \frac{S_1^*}{M_1} = (\sigma_1^1)_\infty \frac{k_{12}+k_{21}}{k_{21}}$$

and

$$(\sigma_1^1)_\infty = \frac{S_1^*}{M_1+M_2}$$

since no isotope is lost from the closed system.

Calculation of Parameters of the Two-Pool System from Values of α's and D's

The values of α_1, α_2, D_{11}, D_{12}, and D_{22} estimated from the specific activity curves can be considered to represent the data obtained in kinetic studies with tracers. Formulas are now presented in which rates, rate constants, pool sizes, and other parameters in the two-pool system are explicitly described in terms of α's and D's. The discussion to follow includes, in addition to the complete system, systems with constraints in which some rates are equal to zero. In these cases, more information is obtained from the same experimental data. Several other situations of special interest, such as those in which two tracers are administered simultaneously, as a rapid injection or as an infusion, are also considered. These results are obtained by minor algebraic rearrangements of preceding equations. When the substitutions are not quite obvious, a number above the equals sign in the formulas indicates the key equation from which the result was derived.

Collection of Formulas for Two-Pool Systems

Case 1. General Case (Fig. 4.5)

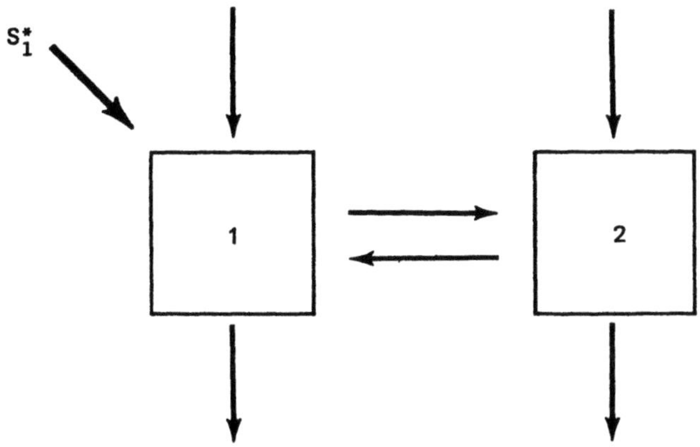

Fig. 4.5. Unrestricted two pool system in which tracer enters through pool 1 (see case 1)

1 a. Only Pool 1 Sampled

$$\sigma_1^1 \overset{4.14}{=} D_{11}\, e^{-\alpha_1 t} + D_{12}\, e^{-\alpha_2 t} \qquad \alpha_1 \neq \alpha_2$$

(if $\alpha_1 = \alpha_2$, the system would have constraints as shown in case 5)

$$M_1 \overset{4.30}{=} \frac{S_1^*}{D_{11} + D_{12}}$$

$$h_{11} \overset{4.2}{=} k_{11} = \frac{r_{11}}{M_1} \overset{4.32}{=} -\frac{\alpha_1 D_{11} + \alpha_2 D_{12}}{D_{11} + D_{12}}$$

$$r_{11} = -(r_{10} + r_{12}) = k_{11} M_1$$

$$h_{22} \overset{4.2}{=} k_{22} = \frac{r_{22}}{M_2} = \frac{-(r_{20} + r_{21})}{M_2} \overset{4.24}{=} -\frac{\alpha_2 D_{11} + \alpha_1 D_{12}}{D_{11} + D_{12}}$$

$$PR_1 \overset{3.2}{=} \frac{S_1^*}{\int_0^\infty \sigma_1^1\, dt} = \frac{S_1^*\, \alpha_1\, \alpha_2}{D_{11}\, \alpha_2 + D_{12}\, \alpha_1}\,.$$

If the concentration of the compound in pool 1 (c_1) is measured, then one can determine the volume of distribution

$$V_1 = \frac{M_1}{c_1}$$

Table 4.1. Parameters of Two-Pool Systems Calculable from Specific Activity Functions [a]

Parameters	Case 1, Fig. 4.5 Unrestricted case Data from pools		Case 2, Fig. 4.6 ($r_{10}=0$) Data from pools		Case 3, Fig. 4.7 ($r_{20}=0$) Data from pools		Case 4, Fig. 4.8 ($r_{01}=0$) Data from pool	Case 5, Fig. 4.9 ($r_{20}=r_{02}=0$) Data from pool	Case 8, Fig. 4.11 (Two tracers) Data from pools	
	1	1 and 2	1	1 and 2	1	1 and 2	1	1	1	1 and 2
M_1	√		√		√		√	√	√	
M_2				√		√		√		√
$h_{11}=k_{11}$	√		√		√		√	√	√	
$h_{22}=k_{22}$	√		√		√		√	√	√	
h_{12}		√		√		√	√	√		√
h_{21}		√		√			√	√		√
k_{12}			√		√			√	√	
k_{21}			√		√			√	√	
k_{10}			(0)		√			√		√
k_{20}			√		(0)			(0)	√	
$-r_{11}=r_{10}+r_{12}$	√		√		√		√	√	√	
$-r_{22}=r_{20}+r_{21}$				√		√				√
r_{12}			√		√			√	√	*
r_{21}		√		√		√	√	√		√*
r_{10}			(0)		√			√	√	*
r_{20}				√	(0)			(0)		√*
r_{01}				√		√	(0)	√		√*
r_{02}				√		√		(0)		√*
PR_1	√*		√*		√*		√*	√*	√*	
PR_2				√*		√		√		√*
ϱ_{12}			(1)			√		√	√*	
ϱ_{21}				√	(1)			(1)		√*
Δ_{12}		√*		√*		√*		(1)		√*
Δ_{21}				√		√	(1)	√		√*

[a] The tracer is administered only into pool 1 in all cases except Case 8 in which two tracers are simultaneously injected, one into pool 1 and the other labeled with a different isotope, into pool 2. The check mark (√) indicates the parameters of the various models, which can be calculated from the specific activity functions determined from either pool 1 or from both pools. The asterisks (*) indicate the parameters which can be calculated from only areas under the curves obtained from the same data. Time parameters are not included but can be estimated as described in Chapter 3 (Table 3.2). Note the similarity of the calculations based on two tracer experiments when either a strict two-pool system (Case 8, Fig. 4.11) or a primary two-pool system embedded in a multicompartmental network (Fig. 1.2; Tables 3.1 and 3.2) are considered.

and the clearance (metabolic clearance rate)

$$MCR_1 = \frac{PR_1}{c_1}$$

1 b. Both Pool 1 and Pool 2 Sampled

If pool 2 is also sampled $[\sigma_1^1 \overset{4.15,\ 4.36}{=} -D_{21}(e^{-\alpha_1 t} - e^{-\alpha_2 t})]$ the following parameters can be estimated in addition to those listed under case 1 a.

$$h_{12} \overset{4.38}{=} -\frac{(\alpha_1 - \alpha_2)\, D_{21}}{D_{11} + D_{12}}$$

$$h_{21} \overset{4.25}{=} \frac{h_{11}\, h_{22} - \alpha_1\, \alpha_2}{h_{12}} \overset{4.20}{=} -\frac{(\alpha_1 - \alpha_2)\, D_{11}\, D_{12}}{D_{21}\,(D_{11} + D_{12})}$$

$$r_{21} = h_{21}\, M_1$$

$$r_{01} = -r_{11} - r_{21}$$

and

$$\Delta_{12} \overset{1.15}{=} \frac{r_{12}}{r_{02} + r_{12}} = \frac{\int_0^\infty \sigma_2^1\, dt}{\int_0^\infty \sigma_1^1\, dt} = \frac{D_{22}\,(\alpha_1 - \alpha_2)}{\alpha_2\, D_{11} + \alpha_1\, D_{12}}\,.$$

Note that M_2, r_{12}, r_{21}, and r_{20} in a two-pool system with no constraints are not calculable from data obtained from experiments in which a tracer is injected into pool 1 only (Table 4.1).

System with Constraints (Some Rates Equal Zero)

Case 2. $r_{10} = 0$ *(Fig. 4.6)*

2 a. Only Pool 1 Sampled

All parameters listed in the general case (1 a) plus the following:

$$r_{12} = -r_{11}$$

$$k_{12} = \frac{r_{12}}{M_1}$$

$$k_{21} \overset{4.25}{=} \frac{k_{11}\, k_{22} - \alpha_1\, \alpha_2}{k_{12}}$$

$$k_{20} = -k_{22} - k_{21}$$

2 b. Both Pool 1 and Pool 2 Sampled

All parameters listed under cases 1 a, 1 b, and 2 a plus the following:

$$M_2 \overset{4.2}{=} \frac{k_{12}\, M_1}{h_{12}}$$

$$r_{21} = k_{21}\, M_2$$

$$r_{20} = k_{20} M_2$$

$$r_{02} = r_{20} + r_{21} - r_{12}$$

$$\varrho_{12} = \frac{r_{12}}{-r_{11}} = 1$$

$$PR_2 = r_{02} + \varrho_{12} r_{01} = r_{02} + r_{01}$$

$$\varrho_{21} = \frac{r_{21}}{r_{20} + r_{21}}$$

$$\Delta_{21} = \frac{r_{21}}{-r_{11}} = \frac{r_{21}}{r_{12}} .$$

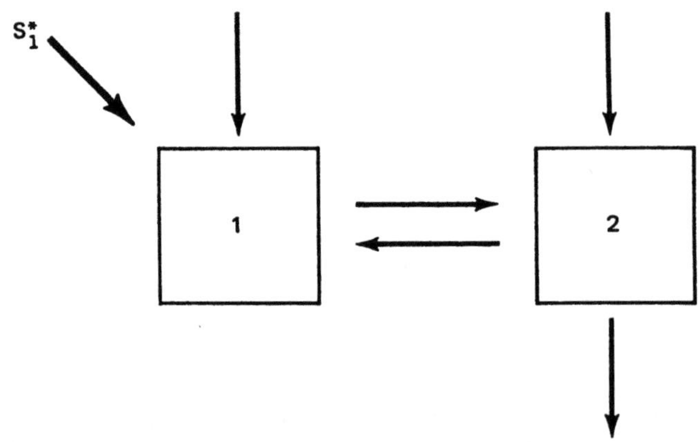

Fig. 4.6. Two pool system in which $r_{10} = 0$ and tracer enters through pool 1 (see case 2)

Note that all the parameters of a two-pool system in which $r_{10} = 0$ can be calculated from data obtained from experiments in which the tracer is injected into pool 1 and the specific activity functions in both pools are determined (Table 4.1).

Case 3. $r_{20} = 0$ (Fig. 4.7)

3 a. Only Pool 1 Sampled

All parameters listed in the general case (1 a) plus the following:

$$k_{21} = -k_{22} = -h_{22}$$

$$k_{12} \overset{4.25}{=} \frac{k_{11} k_{22} - \alpha_1 \alpha_2}{k_{21}}$$

$$r_{12} = k_{12} M_1$$

$$r_{10} = -r_{11} - r_{12}$$

$$k_{10} = \frac{r_{10}}{M_1} = \frac{\alpha_1 \alpha_2}{k_{21}} .$$

3 b. Both Pool 1 and Pool a Sampled

All parameters listed under cases 1 a, 1 b, and 3 a plus the following:

$$M_2 = \frac{k_{12} M_1}{h_{12}} \overset{4.38}{=} \frac{S_1^*}{D_{21}(\alpha_1 - \alpha_2)} \left[\frac{\alpha_1 \alpha_2 (D_{11} + D_{12})}{\alpha_2 D_{11} + \alpha_1 D_{12}} - \frac{\alpha_1 D_{11} + \alpha_2 D_{12}}{D_{11} + D_{12}} \right]$$

$$r_{21} = k_{21} M_2$$

$$r_{02} = r_{12} - r_{21}$$

$$-r_{22} = r_{12} + r_{02}$$

$$\varrho_{12} = \frac{r_{12}}{r_{10} + r_{12}}$$

$$\varrho_{21} = 1$$

$$\Delta_{21} = \frac{r_{21}}{r_{01} + r_{21}}$$

$$PR_2 = r_{02} + \varrho_{12} r_{01}.$$

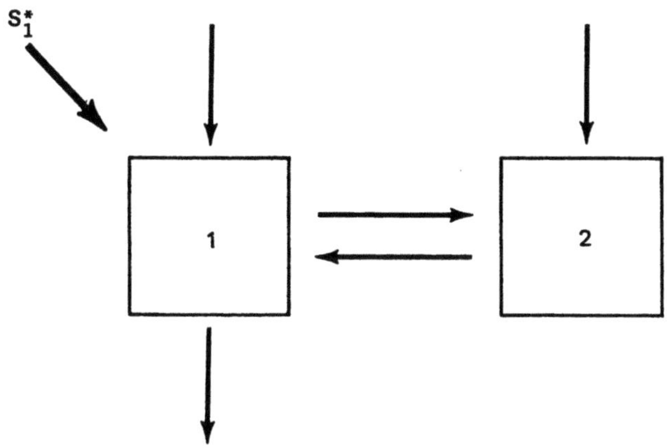

Fig. 4.7. Two pool system in which $r_{20} = 0$ and tracer enters through pool 1 (see case 3)

It can be concluded that all the parameters of two-pool systems in which material is removed irreversibly through only one of the pools ($r_{10} = 0$ or $r_{20} = 0$) can be calculated from the specific activity functions in both pools after administration of tracer into one of the pools (Table 4.1).

If the concentration of the compound in pool 2 (c_2) is measured then $MCR_2 = PR_2/c_2$ and $V_2 = M_2/c_2$ can be determined.

Case 4. $r_{01} = 0$ (Fig. 4.8)

All parameters listed under the general case when only pool 1 is sampled (case 1 a) plus the following:

$$r_{21} = -r_{11}$$

$$h_{21} = \frac{r_{21}}{M_1}$$

$$h_{12} \overset{4.25}{=} \frac{h_{11}\,h_{22} - \alpha_1\,\alpha_2}{h_{21}}$$

$$\Delta_{21} = 1\,.$$

Note that all of these parameters can be calculated from data obtained from pool 1 alone, whereas in the unrestricted case (1 b), samples from pool 2 were necessary to estimate h_{12}, h_{21}, and r_{21}.

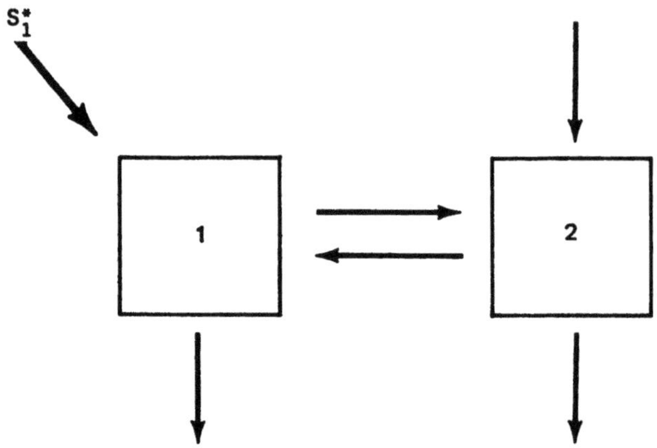

Fig. 4.8. Two pool system in which $r_{01} = 0$ and tracer enters through pool 1 (see case 4)

Case 5. $r_{02} = r_{20} = 0$ (Fig. 4.9)

Only Pool 1 Sampled

All parameters listed under cases 1 a and 3 a plus the following:

$$k_{21} = -k_{22} \qquad k_{12} \overset{4.25}{=} \frac{k_{11}\,k_{22} - \alpha_1\,\alpha_2}{k_{21}}$$

$$r_{12} = r_{21} = k_{12}\,M_1 \qquad h_{21} = \frac{r_{21}}{M_1} = k_{12}$$

$$h_{12} = \frac{r_{12}}{M_2} = k_{21} \qquad M_2 = \frac{r_{21}}{k_{21}} = \frac{k_{12}}{k_{21}}\,M_1$$

$$r_{10} = -r_{11} - r_{12} \qquad r_{01} = r_{10}$$

$$\varrho_{12} = \frac{r_{12}}{-r_{11}} \qquad \varrho_{21} = 1$$

$$\Delta_{12} = 1 \qquad \Delta_{21} = \frac{r_{21}}{-r_{11}}$$

$$PR_1 = r_{01} \qquad PR_2 = \varrho_{12}\,r_{01}\,.$$

Note that all the parameters of this restricted two-pool system can be estimated from the specific activity function of the compound in pool 1 when the tracer is introduced into that pool (Table 4.1). The rewarding complete-

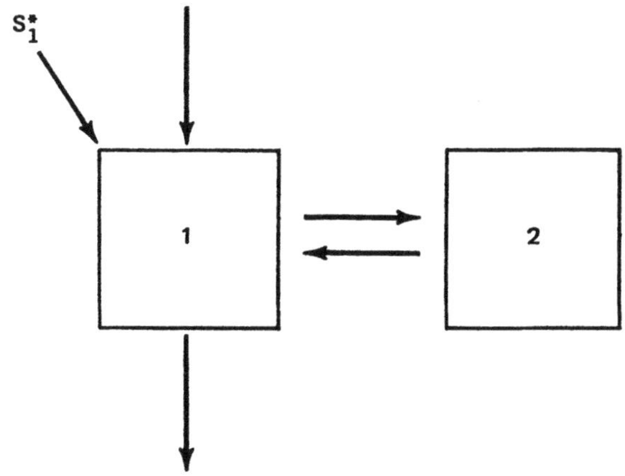

Fig. 4.9. Two pool system in which $r_{02} = r_{20} = 0$ and tracer enters through pool 1 (see case 5)

ness of the calculations that can be made using data from only pool 1 supports, but does not justify, the popularity of this model. For instance, when the tracer of a hormone is injected intravenously and the data on concentration of the labeled compound in circulation can be adequately described by the sum of two exponential terms, an "inner space" (pool 1) and an "outer space" (pool 2) of distribution of the hormone are postulated (TAIT et al., 1961; BRADLEY and WATERHOUSE, 1966). Pool 1 includes the hormone in the vascular system and in tissues, such as liver, into which the hormone is rapidly distributed. Pool 2 includes metabolites or the hormone in other tissues into which it is less rapidly distributed and not metabolized. The volume of the "outer" pool is usually calculated assuming a concentration identical to the plasma level of the hormone.

Case 6. $\alpha_1 = \alpha_2 = \alpha$ (Fig. 4.4)

As described above, this case arises when $h_{11} = h_{22}$ and $h_{21} = 0$. The basic equations are the following:

$$\sigma_1^1 \overset{4.48}{=} (\sigma_1^1)_0 \, e^{-\alpha t}$$

$$\sigma_2^1 \overset{4.49}{=} (\sigma_1^1)_0 \, h_{12} \, t \, e^{-\alpha t}$$

$$h_{11} \overset{4.52}{=} h_{22} = -\alpha$$

$$h_{12} \overset{4.51}{=} \frac{\sigma_2^1}{\sigma_1^1 t}$$

$$M_1 = \frac{S_1^*}{(\sigma_1^1)_0}$$

$$r_{01} = -h_{11} M_1 .$$

Case 7. $r_{10}=r_{20}=r_{01}=r_{02}=0$ (Closed System, Fig. 4.10)

As described in a preceding section, the basic equation in this case is the following:

$$\frac{\sigma_1^1-(\sigma_1^1)_\infty}{(\sigma_1^1)_\infty}\overset{4.56}{=}\frac{k_{12}}{k_{21}}e^{-\alpha t}.$$

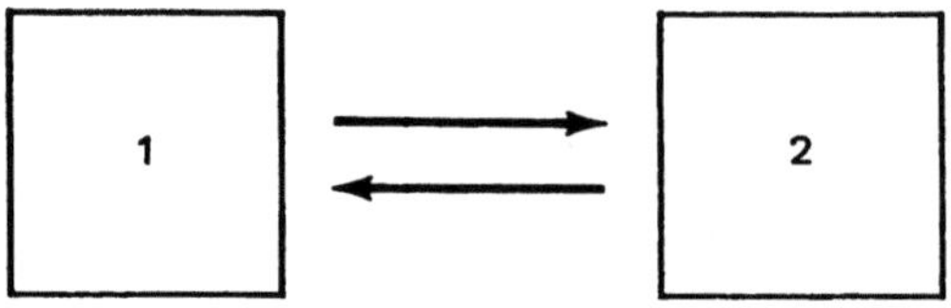

Fig. 4.10. Closed two-pool system (see case 7)

A semilogarithmic plot of $[\sigma_1^1-(\sigma_1^1)_\infty]/(\sigma_1^1)_\infty$ versus time allows the graphic estimation of k_{12}/k_{21} and α. The rate constants k_{12} and k_{21} can be calculated from these data since $\alpha=k_{12}+k_{21}$. From these values, the following parameters can be determined:

$$h_{21}=k_{12}$$

$$h_{12}=k_{21}$$

$$M_1\overset{4.55}{=}\frac{S_1^*k_{21}}{\alpha(\sigma_1^1)_\infty}$$

$$M_2=\frac{S_1^*}{(\sigma_1^1)_\infty}-M_1$$

$$r_{12}=r_{21}=k_{12}M_1.$$

Tracers Injected into Both Pools

Case 8. General Case (Fig. 4.11)

Only Pool 1 Sampled

From the symmetry of the system it follows that, by similarity to Eq. 4.15,

$$\sigma_1^2=D_{12}^{(2)}e^{-\alpha_1 t}+D_{11}^{(2)}e^{-\alpha_2 t}=D_{12}^{(2)}(e^{-\alpha_1 t}-e^{-\alpha_2 t}) \quad (4.56)$$

and, by similarity to Eq. 4.38,

$$D_{12}^{(2)}=-(\sigma_2^2)_0\frac{h_{21}}{\alpha_1-\alpha_2}. \quad (4.57)$$

The superscript (2) attached to the D's serves to identify the coefficients of the equations corresponding to the isotope introduced into pool 2.

At large values of t after the injection, the ratio σ_1^1/σ_1^2 becomes constant since $e^{-\alpha_1 t}$ becomes negligible in comparison to $e^{-\alpha_2 t}$.

Therefore, dividing Eq. 4.14 by 4.55 and using the expressions 4.34 and 4.57, it follows that

$$\left(\frac{\sigma_1^1}{\sigma_1^2}\right)_{t\ \text{large}}\simeq\frac{D_{12}}{-D_{12}^{(2)}}=\frac{(\sigma_1^1)_0(\alpha_1+h_{11})}{(\sigma_2^2)_0\,h_{21}}.$$

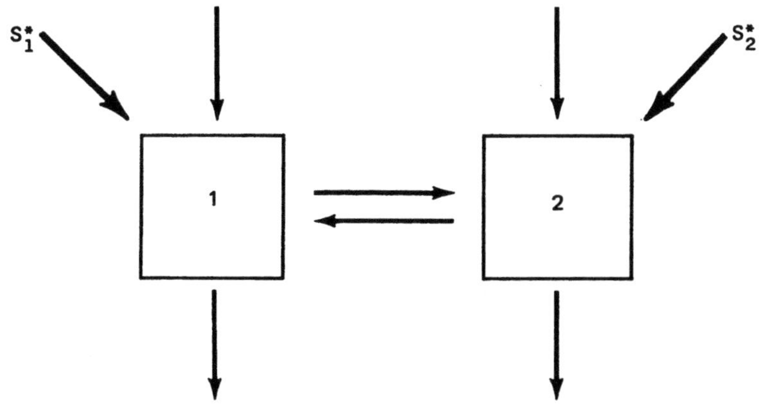

Fig. 4.11. Unrestricted two-pool system in which two different tracers enter through pools 1 and 2 (see case 8)

Since $(\sigma_1^1)_0/(\sigma_2^2)_0 = (S_1^*/M_1)/(S_2^*/M_2)$, $k_{21} = (M_1/M_2)\,h_{21}$, and $h_{11} = k_{11}$, the expression above can be written in the following form:

$$\left(\frac{\sigma_1^1}{\sigma_1^2}\right)_{t\,\mathrm{large}} \cong \frac{S_1^*}{S_2^*}\,\frac{(\alpha_1 + k_{11})}{k_{21}}$$

or

$$k_{21} \cong \frac{S_1^*/S_2^*}{(\sigma_1^1/\sigma_1^2)_{t\,\mathrm{large}}}\,(\alpha_1 + k_{11})\,.$$

Consequently, in addition to all the parameters listed under case 1 a, k_{21} and the following parameters of the system can be calculated from data obtained from pool 1:

$$k_{12} \overset{4.25}{=} \frac{k_{11}\,k_{22} - \alpha_1\,\alpha_2}{k_{21}}$$

$$r_{12} = k_{12}\,M_1$$

$$r_{10} = -r_{11} - r_{12}$$

$$k_{20} = -k_{22} - k_{21}$$

$$\varrho_{12} = \frac{r_{12}}{-r_{11}}\,.$$

It is apparent from the symmetry of the model that the remaining parameters of the system can be calculated from the specific activity functions of the compound in pool 2 (Table 4.1).

An experimental design based on this model has been used in the study of the interconversion of dehydroisoandrosterone and its sulfate (Gurpide *et al.*, 1964; Sandberg *et al.*, 1964). In these experiments advantage was also taken of the simultaneous use of a "metabolically labile isotope" (^{35}S in dehydroisoandrosterone sulfate), which allows the calculation of α_1 from experimental values corresponding to the "tail" of the specific activity curves.

Relationships Between the Specific Activities in Pools 1 and 2

The value of time following the injection of a tracer into pool 1 at which the specific activity of the compound in pool 2 reaches a maximum (t_{max}) can be calculated considering that

$$\left(\frac{d\sigma_2^1}{dt}\right)_{t_{max}} = 0 .$$

It follows from Eq. 4.41 that

$$\frac{d\sigma_2^1}{dt} = \frac{(\sigma_1^1)_0}{\alpha_1 - \alpha_2} h_{12} (\alpha_1 e^{-\alpha_1 t} - \alpha_2 e^{-\alpha_2 t}) \qquad (4.58)$$

and, at t_{max}

$$\alpha_1 e^{-\alpha_1 t_{max}} = \alpha_2 e^{-\alpha_2 t_{max}} .$$

Consequently,

$$e^{(\alpha_1 - \alpha_2) t_{max}} = \frac{\alpha_1}{\alpha_2}$$

and

$$t_{max} = \frac{\ln(\alpha_1/\alpha_2)}{\alpha_1 - \alpha_2} .$$

At t_{max} the same amount of isotope enters and leaves pool 2, i.e.,

$$r_{12} (\sigma_1^1)_{t_{max}} = (r_{20} + r_{21}) (\sigma_2^1)_{t_{max}} .$$

Therefore, if pool 1 is the only precursor of pool 2, i.e., $r_{02} = 0$, then,

$$r_{12} = r_{20} + r_{21}$$

and

$$(\sigma_1^1)_{t_{max}} = (\sigma_2^1)_{t_{max}} .$$

If r_{02} is not zero, then σ_2^1 is smaller than σ_1^1 at t_{max}. These relationships, described by Zilversmit *et al.* (1943), imply that the specific activity curves of pools 1 and 2 intercept at the maximum for pool 2 if 1 is the only precursor of 2.

If Eq. 4.40 is subtracted from Eq. 4.41, it results that

$$\sigma_2^1 - \sigma_1^1 = \frac{(\sigma_2^1)_0}{\alpha_1 - \alpha_2} [(h_{12} - \alpha_1 - h_{11}) e^{-\alpha_2 t} - (h_{12} - \alpha_2 - h_{11}) e^{-\alpha_1 t}] .$$

When $r_{02} = 0$ then $r_{12} = r_{20} + r_{21}$ and $h_{12} = -h_{22}$. Consequently, from Eq. 4.24,

$$h_{12} - \alpha_1 - h_{11} = -h_{22} - \alpha_1 - h_{11} = \alpha_2$$

and

$$h_{12} - \alpha_2 - h_{11} = -h_{22} - \alpha_2 - h_{11} = \alpha_1 .$$

Therefore,

$$\sigma_2^1 - \sigma_1^1 = \frac{(\sigma_1^1)_0}{\alpha_1 - \alpha_2} [\alpha_2 e^{-\alpha_2 t} - \alpha_1 e^{-\alpha_1 t}] \qquad \text{if } r_{02} = 0 .$$

Since σ_2^1 increases with time until t_{max} and decreases afterward, it follows from Eq. 4.58 that $(\alpha_1 e^{-\alpha_1 t} - \alpha_2 e^{-\alpha_2 t})$ is positive for $t < t_{max}$ and negative for

$t > t_{max}$. Therefore, $\sigma_2^1 < \sigma_1^1$ for $t < t_{max}$ and $\sigma_2^1 > \sigma_1^1$ for $t > t_{max}$, when $r_{02} = 0$. ZILVERSMIT *et al.* (1943) also demonstrated very elegantly that the fractional rate h_{12} can be estimated in such systems from the expression

$$h_{12} = \frac{(\sigma_2^1)_{t_2} - (\sigma_2^1)_{t_1}}{\int_{t_1}^{t_2} \sigma_1^1 \, dt - \int_{t_1}^{t_2} \sigma_2^1 \, dt}$$

which can be verified by proper substitutions using Eqs. 4.14, 4.15, 4.24, and 4.39.

Infusion of Tracers into the Pools at a Constant Rate

Using the relation discussed in Chapter 3 (Eq. 3.8) and Eq. 4.40, the functions of time of the specific activity of the compounds in pool 1 during perfusion of the tracer is the following:

$$a_{1\tau}^1 = \frac{I_1^*}{S_1^*} \int_0^\tau \sigma_1^1 \, dt = \frac{I_1^*}{S_1^*} \frac{(\sigma_1^1)_0}{\alpha_1 - \alpha_2} \int_0^\tau [-(\alpha_2 + h_{11}) e^{-\alpha_1 t} + (\alpha_1 + h_{11}) e^{-\alpha_2 t}] \, dt$$

$$= \frac{I_1^*}{M_1 (\alpha_1 - \alpha_2)} \left[\frac{\alpha_2 + h_{11}}{\alpha_1} e^{-\alpha_1 \tau} - \frac{\alpha_1 + h_{11}}{\alpha_2} e^{-\alpha_2 \tau} \right] + a_{1\infty}^1 \qquad (4.59)$$

where

$$a_{1\infty}^1 = \frac{-I_1^*}{M_1 (\alpha_1 - \alpha_2)} \left(\frac{\alpha_2 + h_{11}}{\alpha_1} - \frac{\alpha_1 + h_{11}}{\alpha_2} \right) = \frac{-I_1^* h_{22}}{M_1 \alpha_1 \alpha_2}.$$

Similarly, from Eqs. 3.8 and 4.41,

$$a_{2\tau}^1 = \frac{I_1^*}{S_1^*} \int_0^\tau \sigma_2^1 \, dt = \frac{I_1^*}{S_1^*} (\sigma_1^1)_0 \frac{h_{12}}{\alpha_1 - \alpha_2} \int_0^\tau (-e^{-\alpha_1 t} + e^{-\alpha_2 t}) \, dt$$

$$= \frac{I_1^*}{M_1} \frac{h_{12}}{(\alpha_1 - \alpha_2)} \left(\frac{e^{-\alpha_1 \tau}}{\alpha_1} - \frac{e^{-\alpha_2 \tau}}{\alpha_2} \right) + a_{2\infty}^1 \qquad (4.60)$$

where

$$a_{2\infty}^1 = \frac{-I_1^*}{M_1} \frac{h_{12}}{\alpha_1 - \alpha_2} \left(\frac{1}{\alpha_1} - \frac{1}{\alpha_2} \right) = \frac{I_1^* h_{12}}{M_1 \alpha_1 \alpha_2}.$$

Equivalent equations can be derived to express the specific activities during perfusion of a tracer into pool 2. Thus,

$$a_{1\tau}^2 = \frac{I_2^*}{M_2} \frac{h_{21}}{\alpha_1 - \alpha_2} \left(\frac{e^{-\alpha_1 \tau}}{\alpha_1} - \frac{e^{-\alpha_2 \tau}}{\alpha_2} \right) + a_{2\infty}^1 \qquad (4.61)$$

where

$$a_{1\infty}^2 = \frac{-I_2^*}{M_2} \frac{h_{21}}{\alpha_1 - \alpha_2} \left(\frac{1}{\alpha_1} - \frac{1}{\alpha_2} \right) = \frac{I_2^* h_{21}}{M_1 \alpha_1 \alpha_2}$$

and

$$a_{2\tau}^2 = \frac{I_2^*}{M_2 (\alpha_1 - \alpha_2)} \left(\frac{\alpha_2 + h_{22}}{\alpha_1} e^{-\alpha_1 \tau} - \frac{\alpha_1 + h_{22}}{\alpha_2} e^{-\alpha_2 \tau} \right) + a_{2\infty}^2 \qquad (4.62)$$

where

$$a_{2\infty}^2 = \frac{-I_2^*}{M_2(\alpha_1-\alpha_2)}\left(\frac{\alpha_2+h_{22}}{\alpha_1}-\frac{\alpha_1+h_{22}}{\alpha_2}\right)=\frac{-I_2^* h_{11}}{M_2\,\alpha_1\,\alpha_2}.$$

The specific activity functions resulting after stopping the infusion of the tracer can be calculated as follows. Consider the isotope infused into pool 1. When the perfusion is stopped ($t=0$), the specific activities of the compounds in pools 1 and 2 are $(\sigma_1^1)_0$ and $(\sigma_2^1)_0$. The value of σ_1^1 at any time after stopping the infusion equals the sum of the contributions of the isotope originally present in pool 1 (given by Eq. 4.40) and in pool 2 (given by Eqs. 4.56 and 4.57) when $(\sigma_2^2)_0$ is substituted by $(\sigma_2^1)_0$. Thus,

$$\sigma_1^1=(\sigma_1^1)_0\frac{-(\alpha_2+h_{11})\,e^{-\alpha_1 t}+(\alpha_1+h_{11})\,e^{-\alpha_2 t}}{\alpha_1-\alpha_2}+(\sigma_2^1)_0\frac{-h_{21}}{\alpha_1-\alpha_2}\left(e^{-\alpha_1 t}-e^{-\alpha_2 t}\right). \quad (4.63)$$

Similarly, from Eq. 4.41 and an equation similar to Eq. 4.40 indicating the fate of the isotope present at $t=0$ in pool 2, it follows that

$$\sigma_2^1=(\sigma_1^1)_0\frac{h_{12}\left(-e^{-\alpha_1 t}+e^{-\alpha_2 t}\right)}{\alpha_1-\alpha_2}+(\sigma_2^1)_0\frac{-(\alpha_2+h_{22})\,e^{-\alpha_1 t}+(\alpha_1+h_{22})\,e^{-\alpha_2 t}}{\alpha_1-\alpha_2}. \quad (4.64)$$

C. *Multiple-Pool Systems*

This section describes a general procedure to calculate from tracer data parameters in a system formed by three or more connected pools.

The basic equation that describes the rate of change of the amount of isotope in a pool is Eq. 4.1. An equivalent expression based on specific activity data is given by Eq. 4.3.

The solution of the system of equations 4.3 is

$$\sigma_j^m=\sum_{i=1}^{n} D_{ji}\,e^{-\alpha_i t} \qquad j=1,2,\ldots,n \quad (4.65)$$

when the α's are distinct.

As done when discussing the one- and two-pool systems, it is considered that the D's and the α's are well-determined system parameters, which can be obtained from specific activity data by graphic or computer methods.

The problem of finding expressions which relate the rate constants of the system (the h's in Eq. 4.3) to the calculated values of the D's and the α's is now discussed.

h's in Terms of α's and D's

It follows from Eq. 4.3 $[(\sigma_j^m)'=\Sigma_{i=1}^n h_{ij}\,\sigma_i^m]$ and its successive derivatives evaluated at $t=0$ that

$$(\sigma_j^m)_0^{(p)}=\sum_{i=1}^{n} h_{ij}\,(\sigma_i^m)_0^{(p-1)} \quad (4.66)$$

where (p) indicates the order of the derivative ($p\neq 1$).

It follows from Eq. 4.65 and its derivatives evaluated at $t=0$ that

$$(\sigma_j^m)_0^{(p)} = (-1)^p \sum_{i=1}^{n} \alpha_i^p D_{ji}. \tag{4.67}$$

When $p=0$, no derivative is taken, i.e., $(\sigma_j^m)_0^{(0)} = (\sigma_j^m)_0$. Note that $(\sigma_j^m)_0 = 0$ if $m \neq j$.

All h's can be calculated from the values of D's and α's by solving a system of equations derived from Eqs. 4.66 and 4.67, i.e.,

$$\sum_{i=1}^{n} h_{ij} (\sigma_i^m)_0^{(p-1)} = (-1)^p \sum_{i=1}^{n} \alpha_i^p D_{ji} \qquad i, j = 1, 2, \ldots, n.$$

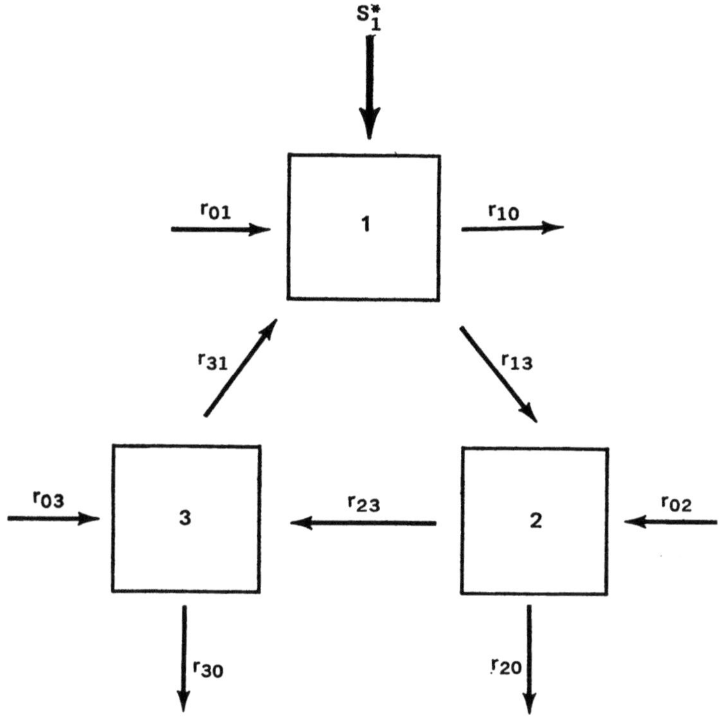

Fig. 4.12. Undirectional cyclic three-pool system

The three-pool model in Fig. 4.12 can be used to illustrate these calculations. The choice of zero values for some rates simplifies the final expression without altering the procedure to calculate h's in terms of D's and α's.

The tracer is introduced into pool 1. According to Eq. 4.66, it follows that, for pool 1 ($j=1$),

$$p=1: \quad (\sigma_1^1)_0' = h_{11} (\sigma_1^1)_0$$

$$p=2: \quad (\sigma_1^1)_0'' = h_{11} (\sigma_1^1)_0' + h_{31} (\sigma_3^1)_0'$$

where the values of $(\sigma_1^1)_0$, $(\sigma_1^1)_0'$, $(\sigma_1^1)_0''$, and $(\sigma_3^1)_0'$ are obtained from Eq. 4.67, i.e.,

$$p=0: \quad (\sigma_1^1)_0 = D_{11}+D_{12}+D_{13}$$

$$p=1: \quad (\sigma_1^1)_0' = -(\alpha_1 D_{11}+\alpha_2 D_{12}+\alpha_3 D_{13})$$

$$(\sigma_3^1)_0' = -(\alpha_1 D_{31}+\alpha_2 D_{32}+\alpha_3 D_{33})$$

$$p=2: \quad (\sigma_1^1)_0'' = \alpha_1^2 D_{11}+\alpha_2^2 D_{12}+\alpha_3^2 D_{13}\,.$$

It is then apparent that h_{11} can be calculated from data obtained from pool 1 alone and h_{31} from data obtained from pools 1 and 3, i.e.,

$$h_{11}=\frac{(\sigma_1^1)_0'}{(\sigma_1^1)_0}=-\frac{\alpha_1 D_{11}+\alpha_2 D_{12}+\alpha_3 D_{13}}{D_{11}+D_{12}+D_{13}}$$

and

$$h_{31}=\frac{(\sigma_1^1)_0''-h_{11}(\sigma_1^1)_0'}{(\sigma_3^1)_0'}=\frac{(\sigma_1^1)_0(\sigma_1^1)_0''-[(\sigma_1^1)_0']^2}{(\sigma_1^1)_0(\sigma_3^1)_0'}\,.$$

Consider now pool 2 ($j=2$). In this case, from Eq. 4.66

$$p=1: \quad (\sigma_1^1)_0' = h_{12}(\sigma_1^1)_0$$

$$p=2: \quad (\sigma_2^1)_0'' = h_{12}(\sigma_1^1)_0' + h_{22}(\sigma_2^1)_0'$$

and, from Eq. 4.67,

$$p=1: \quad (\sigma_2^1)_0' = -(\alpha_1 D_{21}+\alpha_2 D_{22}+\alpha_3 D_{23})$$

$$p=2: \quad (\sigma_2^1)_0'' = \alpha_1^2 D_{21}+\alpha_2^2 D_{22}+\alpha_3^2 D_{23}\,.$$

Therefore, h_{12} and h_{22} can be calculated from data obtained from pools 1 and 3.

A similar analysis applied to pool 3 ($j=3$) gives formulas to calculate h_{33} and h_{23} from data obtained from pools 2 and 3 after injection of the tracer into pool 1.

The size of pool 1 (M_1) is given by the expression

$$M_1=\frac{S_1^*}{(\sigma_1^1)_0}=\frac{S_1^*}{D_{11}+D_{12}+D_{13}}\,.$$

Therefore,

$$r_{11}=h_{11}M_1$$

and

$$r_{31}=h_{31}M_1\,.$$

The values of M_2 and M_3, and therefore of all of the remaining r's in the system depicted in Fig. 4.12, are obtained by injecting tracers into pools 2 and 3 and measuring specific activities in these pools.

Unrestricted Three-Pool Systems (Fig. 4.13)

The specific activity functions of a compound in each of the pools of an unrestricted three-pool system (Fig. 4.13) can be calculated in terms of the

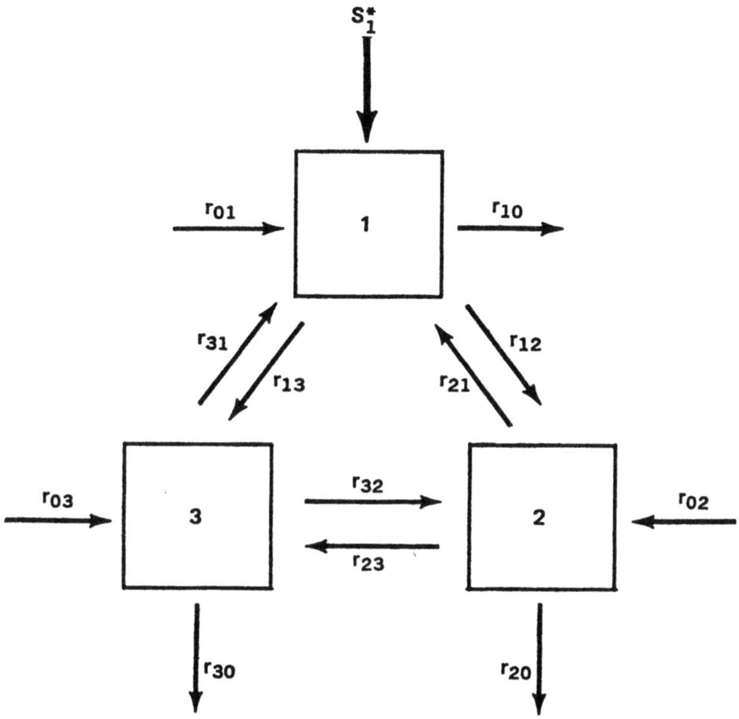

Fig. 4.13. Unrestricted three-pool system

rate constants of the systems, as described above. Complete solutions of this system can be found in the literature (SKINNER *et al.*, 1959). The formulas are given below for reference purpose, as several special cases of this model are often used, particularly in studies of drug metabolism (RESCIGNO and SEGRE, 1966; WAGNER, 1968).

The solutions are given both in terms of rate constants (k's) and fractional rates (h's), parameters which are related by Eq. 4.2.

The specific activities of each of the three pools are the following (Eq. 4.65)

$$\sigma_1^1 = D_{11}\, e^{-\alpha_1 t} + D_{12}\, e^{-\alpha_2 t} + D_{13}\, e^{-\alpha_3 t}$$
$$\sigma_2^1 = D_{21}\, e^{-\alpha_1 t} + D_{22}\, e^{-\alpha_2 t} + D_{23}\, e^{-\alpha_3 t}$$
$$\sigma_3^1 = D_{31}\, e^{-\alpha_1 t} + D_{32}\, e^{-\alpha_2 t} + D_{33}\, e^{-\alpha_3 t}\,.$$

The values of α_1, α_2, and α_3 can be obtained from the following set of equations:

$$\alpha_1 + \alpha_2 + \alpha_3 = -(h_{11} + h_{22} + h_{33}) = -(k_{11} + k_{22} + k_{33})$$
$$\alpha_1\alpha_2 + \alpha_1\alpha_3 + \alpha_2\alpha_3 = h_{11}h_{22} + h_{11}h_{33} + h_{22}h_{33} - h_{12}h_{21} - h_{23}h_{32} - h_{13}h_{31}$$
$$= k_{11}k_{22} + k_{11}k_{33} + k_{22}k_{33} - k_{12}k_{21} - k_{23}k_{32} - k_{13}k_{31}$$
$$\alpha_1\alpha_2\alpha_3 = -h_{11}h_{22}h_{33} + h_{11}h_{23}h_{32} + h_{22}h_{13}h_{31} + h_{33}h_{12}h_{21}$$
$$- h_{13}h_{32}h_{21} - h_{12}h_{23}h_{31}$$
$$= -k_{11}k_{22}k_{33} + k_{11}k_{23}k_{32} + k_{22}k_{13}k_{31} + k_{33}k_{12}k_{21}$$
$$- k_{13}k_{32}k_{21} - k_{12}k_{23}k_{31}\,.$$

The values of the coefficients (D's) are given by the following formulas:

$$D_{11}=(\sigma_1^1)_0\frac{(\alpha_1+h_{22})(\alpha_1+h_{33})-h_{23}h_{32}}{(\alpha_1-\alpha_2)(\alpha_1-\alpha_3)}=(\sigma_1^1)_0\frac{(\alpha_1+k_{22})(\alpha_1+k_{33})-k_{23}k_{32}}{(\alpha_1-\alpha_2)(\alpha_1-\alpha_3)}$$

$$D_{12}=(\sigma_1^1)_0\frac{(\alpha_2+h_{22})(\alpha_2+h_{33})-h_{23}h_{32}}{(\alpha_2-\alpha_3)(\alpha_2-\alpha_1)}=(\sigma_1^1)_0\frac{(\alpha_2+k_{22})(\alpha_2+k_{33})-k_{23}k_{32}}{(\alpha_2-\alpha_3)(\alpha_2-\alpha_1)}$$

$$D_{13}=(\sigma_1^1)_0\frac{(\alpha_3+h_{22})(\alpha_3+h_{33})-h_{23}h_{32}}{(\alpha_3-\alpha_1)(\alpha_3-\alpha_2)}=(\sigma_1^1)_0\frac{(\alpha_3+k_{22})(\alpha_3+k_{33})-k_{23}k_{32}}{(\alpha_3-\alpha_1)(\alpha_3-\alpha_2)}$$

$$D_{21}=(\sigma_1^1)_0\frac{-(\alpha_1+h_{33})h_{12}+h_{13}h_{32}}{(\alpha_1-\alpha_2)(\alpha_1-\alpha_3)}=\left[(\sigma_1^1)_0\frac{M_1}{M_2}\right]\frac{-(\alpha_1+k_{33})k_{12}+k_{13}k_{32}}{(\alpha_1-\alpha_2)(\alpha_1-\alpha_3)}$$

$$D_{22}=(\sigma_1^1)_0\frac{-(\alpha_2+h_{33})h_{12}+h_{13}h_{32}}{(\alpha_2-\alpha_3)(\alpha_2-\alpha_1)}=\left[(\sigma_1^1)_0\frac{M_1}{M_2}\right]\frac{-(\alpha_2+k_{33})k_{12}+k_{13}k_{32}}{(\alpha_2-\alpha_3)(\alpha_2-\alpha_1)}$$

$$D_{23}=(\sigma_1^1)_0\frac{-(\alpha_3+h_{33})h_{12}+h_{13}h_{32}}{(\alpha_3-\alpha_1)(\alpha_3-\alpha_2)}=\left[(\sigma_1^1)_0\frac{M_1}{M_2}\right]\frac{-(\alpha_3+k_{33})k_{12}+k_{13}k_{32}}{(\alpha_3-\alpha_1)(\alpha_3-\alpha_2)}$$

$$D_{31}=(\sigma_1^1)_0\frac{-(\alpha_1+h_{22})h_{13}+h_{12}h_{23}}{(\alpha_1-\alpha_2)(\alpha_1-\alpha_3)}=\left[(\sigma_1^1)_0\frac{M_1}{M_3}\right]\frac{-(\alpha_1+k_{22})k_{13}+k_{12}k_{23}}{(\alpha_1-\alpha_2)(\alpha_1-\alpha_3)}$$

$$D_{32}=(\sigma_1^1)_0\frac{-(\alpha_2+h_{22})h_{13}+h_{12}h_{23}}{(\alpha_2-\alpha_3)(\alpha_2-\alpha_1)}=\left[(\sigma_1^1)_0\frac{M_1}{M_3}\right]\frac{-(\alpha_2+k_{22})k_{13}+k_{12}k_{23}}{(\alpha_2-\alpha_3)(\alpha_2-\alpha_1)}$$

$$D_{33}=(\sigma_1^1)_0\frac{-(\alpha_3+h_{33})h_{13}+h_{12}h_{23}}{(\alpha_3-\alpha_1)(\alpha_3-\alpha_2)}=\left[(\sigma_1^1)_0\frac{M_1}{M_3}\right]\frac{-(\alpha_3+k_{33})k_{13}+k_{12}k_{23}}{(\alpha_3-\alpha_1)(\alpha_3-\alpha_2)}.$$

Solutions for a four-pool unrestricted system have been published by SHIPLEY and CLARK (1972).

Irreducible System: Number of Pools Versus Number of Exponential Terms in Specific Activity Functions

The number of exponential terms in the specific activity function of a pool into which a tracer is introduced does not exceed the total number of pools in an irreducible system to which that pool belongs. However, the number of terms may be much smaller than the number of pools due to particular relations among the rate constants of the system. Note that absolute precision in the determination of specific activity functions is postulated. Therefore, the reduction in the number of terms is a consequence of the structure of the systems and not of experimental inaccuracies.

The essential lack of correlation between the number of pools and exponential terms has been recognized by many authors, and criteria for lumping pools have been described (e.g., SHARNEY *et al.*, 1964; SHIPLEY and CLARK, 1972). Conditions that are *necessary* and *sufficient* to yield a two-exponential specific activity (Eq. 4.14) function in a pool receiving the tracer and exchanging isotope with n—1 other pools have been reported by MANN and GURPIDE

(1969 b) and extended by HEARON (1969) to include more general initial isotope contents in the pools. These conditions are the following:

$$\frac{\sum_{i=2}^{n}\sum_{j=2}^{n} K_{1i}^{(q)} k_{ij} k_{j1}}{\sum_{i=2}^{n} K_{1i}^{(q)} k_{i1}} = \text{constant} \tag{4.68}$$

for all values of q, from $q=1$ to $q=n-1$ and where

$$K_{1i}^{(q)} = \sum_{p_1=1}^{n} \cdots \sum_{p_{q-1}=1}^{n} k_{1p_1} k_{p_1p_2} \cdots k_{p_{q-1}i}$$

and

$$K_{1i}^{(1)} = k_{1i}\,.$$

The symbol $K_{1i}^{(q)}$ is used to simplify the expression in Eq. 4.68 and, as is shown above, indicates the sum of products of many k's. Actually, it includes the sum of all the products of the k's involved in each of the possible pathways of transfer from pool 1 to pool i through the intermediacy of $q-1$ pools.

The same condition, expressed in terms of h's is the following:

$$\frac{\sum_{i=2}^{n}\sum_{j=2}^{n} h_{ij} h_{j1} H_{1i}^{(q)}}{\sum_{i=2}^{n} h_{i1} H_{1i}^{(q)}} = \text{constant} \qquad q=1,2,\ldots,n-1 \tag{4.69}$$

where

$$H_{1i}^{(q)} = \sum_{p_1=1}^{n} \cdots \sum_{p_{q-1}=1}^{n} h_{1p_1} h_{p_1p_2} \cdots h_{p_{q-1}i}$$

and

$$H_{1i}^{(1)} = h_{1i}\,.$$

The value of the constant can be estimated by considering a strict two pool system. Thus, when $n=2$, $i=j=2$, $q=1$, and Eq. 4.69 is reduced to the expression

$$\frac{h_{22} h_{21} h_{1i}}{h_{21} h_{1i}} = \text{constant} = h_{22} = -\frac{D_{11}\,\alpha_2 + D_{12}\,\alpha_1}{D_{11}+D_{12}}\,. \tag{4.70}$$

From this example, and from general considerations (MANN and GURPIDE, 1969 b) it follows that the value of the constant equals the expression $-(D_{11}\,\alpha_2 + D_{12}\,\alpha_1)/(D_{11}+D_{12})$ involving the parameters that define the specific activity function of the compound in pool 1 (Eq. 4.14).

Equations 4.68 and 4.69 indicate that the specific activity function of the compound in the pool receiving the tracer has two exponential terms only when there is a consistent quantitative pattern in the various pathways of transfer of material between the pools. For instance, if the total number of pools in the system is 3 ($n=3$), the necessary and sufficient conditions leading to the function $\sigma_1^1 = D_{11}\,e^{-\alpha_1 t} + D_{12}\,e^{-\alpha_2 t}$ are

$$\frac{k_{12}(k_{22}k_{21}+k_{23}k_{31})+k_{13}(k_{32}k_{21}+k_{33}k_{31})}{k_{12}k_{21}+k_{13}k_{31}}$$

$$= \frac{K_{12}^{(2)}(k_{22}k_{21}+k_{23}k_{31})+K_{13}^{(2)}(k_{32}k_{21}+k_{33}k_{31})}{K_{12}^{(2)}k_{21}+K_{13}^{(2)}k_{31}} \tag{4.70}$$

where

$$K_{12}^{(2)} = k_{11}k_{12}+k_{12}k_{22}+k_{13}k_{32} \quad \text{and} \quad K_{13}^{(2)} = k_{11}k_{13}+k_{12}k_{23}+k_{13}k_{33}\,.$$

The unpleasant complication of these expressions, even in the simple case of three pools and two exponential terms in the experimental curve, defies attempts to gain a complete intuitive grasp of the conditions. However, Eq. 4.68 is useful in that it provides the means to search for models and conditions that are compatible with an experimental curve described by the sum of two exponential terms. Some examples of application of this equation to the interpretation of isotopic data obtained from blood-borne compounds are given in Chapter 5.

Linearly Dependent Specific Activities

It has been shown (HEARON, 1963; MANN and GURPIDE, 1969 a) that in an n-pool irreducible system (each of the n pools exchange isotope with all the others) a linear dependence among the specific activities of the pools leads to a reduction in the number of exponential terms. Thus, if

$$\sigma_p^1 = \sum_{j=1}^{m} A_{pj}\, \sigma_j^1 \qquad p = m+1, \ldots, n \tag{4.71}$$

then the general expression

$$\sigma_j^1 = \sum_{i=1}^{n} D_{ji}\, e^{-\alpha_i t} \tag{4.72}$$

is reduced to

$$\sigma_j^1 = \sum_{i=1}^{m} D_{ji}\, e^{-\alpha_i t} \qquad j = 1, 2, \ldots, m \tag{4.73}$$

where the α's are renumbered if necessary so that the missing terms correspond to the last $n-m$ terms of Eq. 4.72.

Furthermore, if each σ_j^1 in the right-hand side of Eq. 4.71 is replaced by the sum of m exponential terms indicated by Eq. 4.73, it follows that σ_p^1 with $p > m$, is also equal to a linear combination of the same m exponential terms, i.e.,

$$\sigma_j^1 = \sum_{i=1}^{m} D_{ji}\, e^{-\alpha_i t} \qquad j = 1, 2, \ldots, n\,. \tag{4.74}$$

Conversely, if the specific activities of all pools in the irreducible system (1 to n) can be written as a linear combination of the same m exponential terms, then the specific activity of $n-m$ pools are linear combinations of the specific activities of m pools.

The linear dependence of the specific activities of $n-m$ pools upon the specific activities of m others results from relationships among the rate constants in the system. The necessary and sufficient conditions for the existence of a set of number A_{pj} satisfying Eq. 4.71 are the following (MANN and GURPIDE, 1969 a):

$$h_{ji} + \sum_{q=m+1}^{n} A_{qj}\, h_{qi} = \sum_{p=2}^{m} A_{ip}\left(h_{jp} + \sum_{q=m+1}^{n} A_{qj}\, h_{qp}\right) \qquad i > m,\ j = 1, 2, \ldots, m \tag{4.75}$$

and $$(\sigma_i^{(k)})_0 = \sum_{j=2}^{m} A_{ij}\,(\sigma_j^{(k)})_0 \qquad i > m,\ j = 2, \ldots, m\,. \tag{4.76}$$

If the relation between specific activities is one of direct proportionality, i.e.,

$$A_{ij} = \frac{\sigma_i^1}{\sigma_j^1} \qquad i > m,\ j = 2, \ldots, m$$

then $$A_{ij} = \frac{\sigma_i^1}{\sigma_j^1} = \frac{(\sigma_i^1)'}{(\sigma_j^1)'} = \frac{(\sigma_i^1)_0'}{(\sigma_j^1)_0'} = \frac{h_{1i}}{h_{1j}}\,.$$

If, in addition, the specific activity functions of all the pools in the irreducible system are a sum of two exponential terms with the same exponential constants (α_1 and α_2), then $m = 2$ and Eqs. 4.75 and 4.76 are simplified to the following expressions:

$$h_{1i} + \sum_{q=3}^{n} \frac{h_{1q}}{h_{11}}\,h_{qi} = \frac{h_{1i}}{h_{12}}\left(h_{12} + \sum_{q=3}^{n} \frac{h_{1q}}{h_{11}}\,h_{q2}\right) \qquad i > 2$$

$$h_{2i} + \sum_{q=3}^{n} \frac{h_{1q}}{h_{12}}\,h_{qi} = \frac{h_{1i}}{h_{12}}\left(h_{22} + \sum_{q=3}^{n} \frac{h_{1q}}{h_{12}}\,h_{q2}\right) \qquad i > 2$$

and $$(\sigma_i^{(k)})_0 = \frac{h_{1i}}{h_{12}}\,(\sigma_2^{(k)})_0 \qquad i > 2,\ k < n-2\,.$$

These conditions are summarized in the following expression (GURPIDE and MANN, 1970):

$$H_{li}^{(q)} = \frac{h_{1i}}{h_{12}}\,H_{12}^{(q)} \qquad i > 1,\ q = 1, 2, \ldots, n-3 \tag{4.77}$$

where $H_{li}^{(q)}$ has the meaning given with Eq. 4.69, i.e.,

$$H_{li}^{(q)} = \sum_{p_1=1}^{n} \cdots \sum_{p_{q-1}=1}^{n} h_{1p_1}\,h_{p_1p_2} \cdots h_{p_{q-1}i}$$

and $$H_{li}^{(1)} = h_{1i}\,.$$

The same conditions can be expressed in terms of k's by direct substitution of h_{ij} by k_{ij}.

Pool 2 is one of the peripheral pools exchanging isotope with pool 1 (the pool into which the tracer is introduced) and is taken arbitrarily as a pool of reference.

For instance, in a irreducible three-pool system, $n = 3$ and Eq. 4.77 is simplified to the single expression

$$h_{12}\,h_{23} + h_{13}\,h_{33} = \frac{h_{13}}{h_{12}}\,(h_{12}\,h_{22} + h_{13}\,h_{32}) \tag{4.78}$$

or, in terms of k's,

$$k_{12}\,k_{23} + k_{13}\,k_{33} = \frac{k_{13}}{k_{12}}\,(k_{12}\,k_{22} + k_{13}\,k_{32})\,. \tag{4.79}$$

Since Eq. 4.78 describes a relationship among the h's that is necessary and sufficient to produce double exponential specific activity functions in all pools in the system, it obviously provides a condition that satisfies Eq. 4.70 as well.

These relationships are further discussed in the next chapter where they are applied to the interpretation of specific activity curves of circulating hormones.

Repeated α's

As indicated when describing the two-pool system, $(-\alpha_1)$, $(-\alpha_2)$, ..., $(-\alpha_n)$, in Eq. 4.65 are the roots (eigenvalues) of the equation of n-th degree in α given by the following determinant:

$$\begin{vmatrix} h_{11}-\alpha & h_{21} & \cdots & h_{n1} \\ h_{12} & h_{22}-\alpha & \cdots & h_{n2} \\ \vdots & \vdots & & \\ h_{1n} & h_{2n} & \cdots & h_{nn}-\alpha \end{vmatrix} = 0 \tag{4.80}$$

If some of the α's are repeated, Eq. 4.65 is not the solution of the system of equations 4.3. Instead of a sum of exponential terms the solution is a sum of exponential terms multiplied by t^n, $n = 0, 1, \ldots, m$, where m is the maximum number of repetitions of a root. Thus, if in a four-pool system, three of the roots are identical ($\alpha_1 = \alpha_2 = \alpha_3$) then

$$\sigma_j^m = D_{j1}\, e^{-\alpha_1 t} + D_{j2}\, t e^{-\alpha_1 t} + D_{j3}\, t^2 e^{-\alpha_1 t} + D_{j4}\, e^{-\alpha_4 t}.$$

The existence of a situation in which roots are repeated may be detected from the analysis of specific activity curves.

Complex α's

The roots calculated by solving Eq. 4.80 may be complex and still be perfectly suitable for use in Eq. 4.65. The characteristics of a system containing complex roots are the following:

1. If there is a complex root (e.g., $\alpha_1 = A + Bi$), there is another complex root, which is the conjugate of the first (e.g., $\alpha_2 = A - Bi$) (HEARON, 1963).

2. If α_m is the smallest of all the eigenvalues of the matrix $[h_{ij}]$ (Eq. 4.80) in absolute value, then it is positive, real, and smaller than the real part of any other α (HEARON, 1963). In general, in any irreducible system, semilogarithmic plot of the specific activities of each pool in the system results in parallel "tails," indicating that the specific activities in all pools are practically proportional at values of t sufficiently large.

3. Systems with complex roots can be expected to show oscillations and, in special cases, even a nonmonotonic decline of specific activity values (THRON, 1972). Consider the function

$$\sigma = D_1\, e^{-\alpha_1 t} + D_2\, e^{-\alpha_2 t} + D_3\, e^{-\alpha_3 t}$$

in which α_1 and α_2 are complex, i.e.,

$$\alpha_1 = A + Bi$$
$$\alpha_2 = A - Bi$$

and α_3 is real. Thus,

$$\sigma = D_1\, e^{-(A+Bi)t} + D_2\, e^{-(A-Bi)t} + D_3\, e^{-\alpha_3 t}$$

or

$$\sigma = D_1\, e^{-At}\, e^{-Bit} + D_2\, e^{-At}\, e^{Bit} + D_3\, e^{-\alpha_3 t}.$$

Since $e^{-ix} = \cos x - i \sin x$ and $e^{ix} = \cos x + i \sin x$ (AGNEW, 1960, p. 180), then

$$\sigma = D_1 e^{-At} (\cos Bt - i \sin Bt) + D_2 e^{-At} (\cos Bt + i \sin Bt) + D_3 e^{-\alpha_3 t}$$

or

$$\sigma = e^{-At} [(D_1 + D_2) \cos Bt - (D_1 - D_2) i \sin Bt] + D_3 e^{-\alpha_3 t} .$$

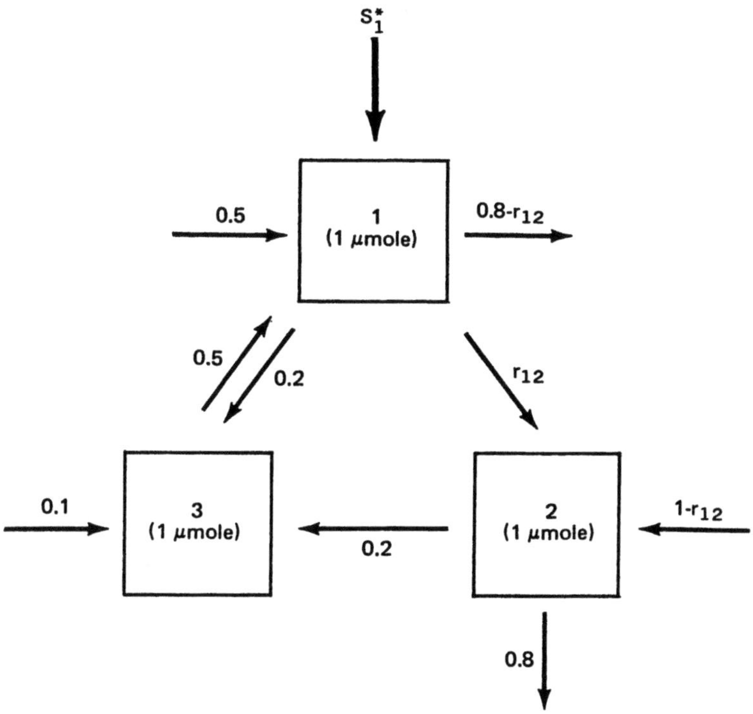

Fig. 4.14. Three-pool system in which the α's are distinct, repeated, or complex according to the value of r_{12}.
If $r_{12} = 0.010$, $\sigma_1 = 86.9\, e^{-1.44t} - 6.0\, e^{-0.010t} + 19.1\, e^{-0.345t}$.
If $r_{12} = 0.043$, $\sigma_1 = (80.6 - 6.3\, t)\, e^{-1.080t} + 19.4\, e^{-1.339t}$.
If $r_{12} = 0.500$, $\sigma_1 = 83.6\, e^{-1.115t} \sin(-0.235\, t + 1.209) + 21.9\, e^{-0.269t}$.
(Example taken from GURPIDE and MANN, 1970.)

Since the system is assumed to be physically realizable, $D_1 - D_2$ must be a pure imaginary number (in order to ensure a real value for the sin Bt term) and $D_1 + D_2$ must be real. Therefore, D_1 and D_2 must be complex conjugates. It then follows that

$$\sigma = e^{-At} (D_4 \cos Bt + D_5 \sin Bt) + D_3 e^{-\alpha_3 t}$$

where $D_4 = D_1 + D_2$, $D_5 = D_1 - D_2$, and D_3 are real. This expression can be written in an equivalent form (DEFARES and SNEDDON, 1961, p. 102):

$$\sigma = D \sin(Bt + C) e^{-At} + D_3 e^{-\alpha_3 t} . \tag{4.81}$$

Equation 4.81 makes explicit the oscillatory character of the exponential function with complex exponents, as well as the dampening effect of the factor e^{-At} on the amplitude of the oscillations. Furthermore, since $\alpha_3 < A$, the term containing the oscillating component will become negligible with respect to $D_3\, e^{-\alpha_3 t}$ as t increases.

Figure 4.14 shows a three-pool system and illustrates numerically how changes in the value of only one rate constant may result in either distinct, repeated, or complex α's.

Chapter 5

Interpretation of Isotopic Data from Blood - Borne Compounds

In a large number of reported *in vivo* tracer studies, labeled hormones are either infused or injected intravenously and concentrations or specific activities of the labeled compounds isolated from blood or plasma are measured. Typically, the isotopic data from these experiments are steady-state concentration values (c^*, in dpm/ml), obtained during infusion of the tracer, or values of concentrations determined in samples drawn at various intervals of time after injection of the tracer. Endogenous concentrations (c, in nmole/ml) of the blood-borne compounds are usually measured before beginning the administration of tracers or, more properly, in each individual sample taken during the experiment. Specific activities (σ, in dpm/nmole) are then calculated by dividing these two quantities ($\sigma = c^*/c$). It should be pointed out that specific activities can also be measured directly (e.g., by double isotope derivative procedures (ULICK *et al.*, 1958), without determining the actual concentration of the compound in blood or plasma. The applicability of the concepts presented in previous chapters to the interpretation of data obtained from this type of experiments is now discussed.

As a rule, a part of any hormone introduced intravenously returns to the vascular space after distribution in other spaces or participation in reversible reactions. Therefore, the kinetics of removal of the tracer from circulation usually cannot be described by single exponential decay curves or by one-pool models and more complex systems must be considered. Most suitable for this purpose are the analyses presented in Chapters 1 to 3 since they are applicable to systems in which a few accesible pools exchange isotope with a large and undetermined number of other pools. In the following discussions, a compound in systemic circulation is considered to represent one pool, although the estimation of some basic parameters (e.g., production rates, metabolic clearance rates) can also be justified using noncompartmental arguments.

The topics treated in this chapter are ordered according to the complexity (and reliability) of measurements and assumptions: the estimation of steady-state production rates by administration of a single tracer is discussed first and the estimation of rates of interconversion of three circulating hormones and non-steady state conditions are considered last.

I. Data: Isotopic Steady-State Values or Areas under Concentration Curves

A. One Tracer Administered

Production Rates

Equation 2.1 can be used to calculate the production rate (PR) of a circulating compound from steady-state data obtained during intravenous infusion of a tracer. Equation 3.2 shows an equivalent expression based on the total area under the specific activity curves obtained when the tracer is injected as a single dose. These calculations imply the assumption that the production rate of a hormone in systemic circulation, a fundamental parameter which represents the rate at which the hormone is supplied to the blood stream, is constant. The applicability of these formulas to a non-steady-state situation is discussed at the end of this chapter.

As indicated in Chapter 2, it is not necessary to assume homogenous distribution of the tracer in circulation to validate the calculations of production rates by the expressions I^*/a or $S^*/\int_0^\infty \sigma \, dt$. These expressions were derived assuming a uniform specific activity rather than uniform concentration of the compound in blood (Eqs. 2.1 and 3.2). Uniform specific activities in the presence of variable local concentrations of the hormone can be expected when tracer and secreted hormone enter the circulation at equivalent sites or when they are rapidly mixed.

The first situation (equivalent sites of entry of tracer and unlabeled hormone) is of common occurrence. A tracer infused into a peripheral vein as well as the secretory products of adrenal, gonadal, thyroidal, or pituitary glands are transported immediately to a common site, viz. the right heart. Catabolic processes undergone by these hormones do not change the specific activity established at that site. In contrast, pancreatic or gastric hormones, which must pass through the liver on their way to the heart, may not be studied properly with a peripherally administered tracer. In this case, the tracer is not exposed to hepatic metabolism to the same extent as the endogenously produced hormone (FIELD, 1972).

The importance of the site of infusion of the tracer can be illustrated with results obtained in a series of experiments in which the metabolism of progesterone by sheep fetuses *in utero* was studied (GURPIDE *et al.*, 1972). Catheters were placed into an umbilical and a fetal femoral vein and blood was sampled from a fetal femoral artery. Tritiated progesterone was infused into the umbilical vein, and, simultaneously, [^{14}C]-labeled hormone was infused through the femoral catheter. At the steady state, the [^{3}H] : [^{14}C] ratio in progesterone isolated from the femoral arterial blood was about 30% lower than the [^{3}H] : [^{14}C] ratio of the infused tracers, indicating that part of the tritiated hormone was destroyed by hepatic metabolism during its transit from the placenta to the inferior vena cava. If the rate of supply of the hormone to the fetus is to be measured, the tracer should be infused into the umbilical vein since progesterone enters the fetal circulation through the umbilical vessels; erroneous results would be obtained if the tracer is infused into a femoral vein.

The second situation leading to uniform specific activity of the hormone throughout the vascular system is characterized by rapid mixing of the tracer with the endogenous circulating hormone. The situation prevails when the removal of the hormone from blood is slow in comparison to the circulation time. Although satisfied for many hormones, uniform mixing may not be achieved with compounds such as estradiol, progesterone, hypothalamic releasing factors, and other hormones that are removed from blood very rapidly. In these cases, a significant portion of the dose may be lost during one cycle through the vascular system.

The period of distribution of the tracer in the vascular system is the subject of detailed studies. ZIERLER (1963) and others have emphasized the dispersion of transit times through capillary beds of a tracer introduced into circulation. WATERHOUSE and KEILSON (1972) also analyzed the process of distribution of a tracer in circulation and presented data indicating that after a fast (1 sec) introduction of a bolus of tracer ([^{14}C]glucose or [^{31}I]albumin) into a peripheral vein the isotope concentration in arterial blood reached a maximum at about 18 sec and declined sharply by the end of the first minute. This departure from the postulated "instantaneous mixing" characterizing an ideal pool is of no significance for the estimations of production rates of hormones but must be kept in mind when evaluating the validity of certain calculations, particularly those involving values at time zero.

Metabolic Clearance Rates

The concept of clearance and its importance in physiology is well known and has been described lucidly by RIGGS (1963) and by STEELE (1971) who indicate its usefulness in the study of renal function.

The parameter usually called "metabolic clearance rate" (MCR) of a circulating hormone (DOMINGUEZ, 1950; ROTHSCHILD *et al.*, 1957; TAIT, 1963; TAIT and BURSTEIN, 1964) can be defined as the ratio of the rate of irreversible removal of the hormone from the bloodstream to its concentration in peripheral blood. MCR is a useful index of metabolism, which can be determined easily from isotopic data without measuring endogenous blood levels of the hormone (Eq. 2.2). Constant MCR values can be noted even under conditions of variable rates of production of the endogenous hormone since the concentrations of the exogenously administered labeled compounds are determined by the fractional rates of metabolism, which are usually unaffected by changes in the production rates of the hormone within the physiologic range. For example, the metabolic clearance rates of progesterone in men and in pregnant and nonpregnant women are very similar in spite of large differences in the rates of secretion of the hormone (LIN *et al.*, 1972). Similarly, the MCR of placental lactogen measured in men or nonpregnant women is not significantly different from the pregnancy value (KAPLAN *et al.*, 1968). In some instances, MCR's of hormones (estradiol, HEMBREE *et al.*, 1969; oxytocin, FABIAN *et al.*, 1968) were determined at widely different rates of infusion (micrograms per hour) and no changes were noted until definitely unphysiologic amounts were administered. There are, however, clear examples of dependence of metabolic clearance rates upon the rate of production of the hormone. The MCR of testosterone, for instance, is affected by changes in its blood level (SOUTHREN *et al.*, 1965, 1968; BARDIN and LIPSETT, 1967). Moreover, the MCR of cortisol measured in the morning, differs from values obtained when the experiments are performed in the afternoon (DE LACERDA *et al.*, 1973). It should also be noted that considerable variations in MCR values have been observed by changing the posture of the subject from the supine to the standing position, likely due to changes in the hepatic blood flow (BALIKIAN *et al.*, 1968; FLOOD *et al.*, 1973).

Extensive series of measurements of MCR of hormones in men and women have been performed, as is indicated in Table 5.1.

Average values obtained from normal subjects are often used to estimate production rates from data on blood levels of the hormone (Eq. 2.2), assuming that a steady-state prevails. Of course, variability of MCR values from individual to individual must be taken into account when evaluating the accuracy of the calculations.

Most MCR values reported in the literature are expressed as liters of plasma per day (or liters of plasma per day per square meter of body surface) because plasma samples are used for the measurements of concentrations of labeled hormones. Plasma, rather than whole blood, is analyzed since, in general, the fraction of the circulating hormone associated with erythrocytes is negligible. This is not always the case; a significant proportion of estrone, for instance, is bound to the red blood cells (LONGCOPE *et al.*, 1968). Since the hormone in plasma or in erythrocytes may be cleared at different rates, as was shown for testosterone (RAUS *et al.*, 1968), metabolic clearance rates determined from plasma or whole blood samples may not be equivalent.

The preceding comments imply that the exchange of hormone between red blood cells and plasma is slow enough to allow for a distinction of their kinetic behavior. In contrast, the exchange between unbound and protein-bound hormone in plasma at 37° C appears to be extremely fast (SANDBERG *et al.*, 1966). Consequently, usual procedures, which involve extraction of the hormone from plasma with solvents and do not distinguish between the bound and unbound species, appear to be justified from a kinetic standpoint. It is important, however, to test for compartmentalization in plasma for each com-

Table 5.1. Metabolic clearance rates of some hormones and metabolites in man

Compound	Subject		Sample peripheral venous:	MCR		Reference
	Sex	Condition		liter/day	liter/day m²	
Estradiol	F		Plasma	1350	790	Longcope *et al.* (1968)
			Blood	2280	1360	Longcope *et al.* (1968)
			Plasma	1025	615	Hembree *et al.* (1969)
		Breast ca	Plasma	766	—	Pearlman *et al.* (1969)
	M		Plasma	1890	990	Longcope *et al.* (1968)
			Blood	3060	1600	Longcope *et al.* (1968)
			Plasma	1640	830	Hembree *et al.* (1969)
			Plasma	—	910	Longcope and Tait (1971)
Estrone	F		Plasma	2210	1320	Longcope *et al.* (1968)
			Blood	3200	1910	Longcope *et al.* (1968)
			Plasma	1630	980	Hembree *et al.* (1969)
		Breast ca	Plasma	2240	—	Pearlman *et al.* (1969)
	M		Plasma	2570	1310	Longcope *et al.* (1968)
			Blood	3880	1870	Longcope *et al.* (1968)
			Plasma	2410	1170	Hembree *et al.* (1969)
			Plasma	—	1170	Longcope and Tait (1971)
Estrone sulfate	F		Plasma	146	94	Ruder *et al.* (1972)
	M		Plasma	113	55	Ruder *et al.* (1972)
			Plasma	200	105	Longcope (1972)
Androstenedione	F and M		Plasma	2200	—	Baird *et al.* (1968) [a]
			Plasma	2100	1160	Longcope *et al.* (1969)
			Plasma	2070	1100	Mahoudeau *et al.* (1971)

Table 5.1. (continued)

Compound	Subject		Sample peripheral venous:	MCR		Reference
	Sex	Condition		liter/day	liter/day m²	
Testosterone	F		Plasma	690	—	Baird *et al.* (1968) [a]
			Plasma	440	275	Longcope *et al.* (1969)
			Plasma	530	—	Bird *et al.* (1971)
			Plasma	—	304	Saez *et al.* (1972)
		Pregnancy	Plasma	—	192	Saez *et al.* (1972)
	M		Plasma	1080	—	Baird *et al.* (1968) [a]
			Plasma	930	465	Longcope *et al.* (1969)
			Plasma	1080	—	Bird *et al.* (1971)
			Plasma	1030	545	Mahoudeau *et al.* (1971)
			Plasma	—	516	Saez *et al.* (1972)
Testosterone sulfate	F	Breast ca	Blood	21—25	—	Wang *et al.* (1967 a)
	M		Plasma	75—146	50—89	Saez *et al.* (1971)
5α-Dihydrotesterone	F		Plasma	314	—	Ito and Horton (1971)
			Plasma	—	209	Saez *et al.* (1972)
		Pregnancy	Plasma	—	89	Saez *et al.* (1972)
			Plasma	243	153	Mahoudeau *et al.* (1971)
	M		Plasma	634	336	Mahoudeau *et al.* (1971)
			Plasma	652	—	Ito and Horton (1971)
			Plasma	—	391	Saez *et al.* (1972)
Progesterone	F		Plasma	2510	1510	Lin *et al.* (1972)
		Ovariectomy	Plasma	2170	1360	Lin *et al.* (1972)
		Pregnancy	Plasma	2020	1150	Lin *et al.* (1972)
	M		Plasma	2080	1180	Lin *et al.* (1972)

Table 5.1. (continued)

Compound	Subject		Sample peripheral venous:	MCR		Reference
	Sex	Condition		liter/day	liter/day m²	
20α-Dihydroprogesterone	F		Plasma	2200	—	Lin *et al.* (1972)
Pregnenolone	F	Pregnancy 11—12 wks	Blood	2950	—	Little *et al.* (1971)
		33—36 wks	Blood	2670	—	Little *et al.* (1971)
		37—39 wks	Blood	4530	2770	Little *et al.* (1971)
	M		Blood	4060	—	Little and Billiar (1969)
			Plasma	3310	—	Little and Billiar (1969)
			Blood	1260—3050	—	Wang *et al.* (1967 a)
Pregnenolone sulfate	M		Blood	280	—	Wang *et al.* (1967 a)
17-Hydroxypregnenolone	F		Plasma	2000	—	Strott *et al.* (1970)
Cortisol	F and M		Plasma	219	122	Segre *et al.* (1966)
			Plasma	116	—	Bradley and Waterhouse (1966)
			Plasma	247	155	Dazord *et al.* (1972)
			Plasma	173—290	91—149	de Lacerda *et al.* (1973)
Cortisone	F and M		Plasma	722	451	Dazord *et al.* (1972)
Deoxycorticosterone	F and M		Plasma	960	—	Schambelan and Biglieri (1972)
Aldosterone	F		Plasma	1620	—	Tait *et al.* (1961)
		Pregnancy	Plasma	1540	—	Tait *et al.* (1962)
		Pregnancy	Blood	2240	—	Tait *et al.* (1962)
	M		Plasma	1300—1990	—	Balikian *et al.* (1968)
	F and M		Plasma	1340	—	Lustscher *et al.* (1963)
			Plasma	1330	—	Ford *et al.* (1968)

Table 5.1. (continued)

Compound	Subject		Sample peripheral venous:	MCR		Reference
	Sex	Condition		liter/day	liter/(day m²)	
Dehydroisoandrosterone	F and M		Plasma	1640	—	Horton and Tait (1967)
	F	Breast ca	Blood	960	—	Wang *et al.* (1967 b)
Dehydroisoandrosterone sulfate	F		Plasma	8	—	Sandberg *et al.* (1964)
		Breast ca	Blood	17—32	—	Wang *et al.* (1967 b)
			Plasma	6.4—8	—	Gant *et al.* (1971)
		Pregnancy	Plasma	55—74	—	Gant *et al.* (1971)
	M		Plasma	6.2—8.4	—	Gant *et al.* (1971)
Cholesterol sulfate	M		Plasma	11—18	—	Gurpide *et al.* (1966 a)
Thyroxine	F and M		Serum	1.1	—	Oppenheimer *et al.* (1970)
			Serum	1.2	—	Koutras *et al* (1970)
			Serum	1.3	—	Nicoloff *et al.* (1972)
		Endemic goiter	Serum	0.54	—	Koutras *et al.* (1970)
Triiodothyronine	F and M		Serum	25	—	Oppenheimer *et al.* (1970)
			Serum	22	—	Woeber *et al.* (1970)
			Serum	26	—	Nicoloff *et al.* (1972)
		Hypothyroidism	Serum	27	—	Nicoloff *et al.* (1972)
		Hyperthyroidism	Serum	74	—	Nicoloff *et al.* (1972)
		Endemic goiter		20	—	Koutras *et al.* (1970)
	M		Serum	26	—	Cavalieri *et al.* (1971)
		Hyperthyroidism	Serum	52—79	—	Cavalieri *et al.* (1971)
Diiodotyrosine	F and M		Serum	111	—	Singer and Nicoloff (1972)
TSH	F and M		Plasma	47	—	Becker and Cornette (1971)
			Plasma	61	—	Odell *et al.* (1967)
		Hypothyroidism	Plasma	52	—	Odell *et al.* (1967)

Table 5.1. (continued)

Compound	Subject: Sex	Subject: Condition	Sample peripheral venous	MCR: liter/day	MCR: liter/(day m²)	Reference
FSH	F		Plasma	20	—	Coble *et al.* 1969)
		Postmenopause	Plasma	18	—	Coble *et al.* (1969)
LH	F		Plasma	27	—	Coble *et al.* (1969)
			Plasma	35	—	Kohler *et al.* (1968)
		Postmenopause	Plasma	37	—	Kohler *et al.* (1968)
hCG	F		Plasma	3.4	—	Rizkallah *et al.* (1969)
	M		Plasma	3.9	—	Rizkallah *et al.* (1969)
hPL	F		Plasma	184	—	Kaplan *et al.* (1968)
		Pregnancy	Plasma	170	—	Kaplan *et al.* (1968)
	M		Plasma	152	—	Kaplan *et al.* (1968)
hGH	F and M		Plasma		203	Cameron *et al.* (1969)
			Plasma	330	180	Taylor *et al.* (1969)
			Plasma	—	160	MacGillivray *et al.* (1970)
			Plasma	330—480	—	Kowarski *et al.* (1971)
			Plasma	160—300	—	Thompson *et al.* (1972)
		Acromegaly	Plasma	—	170	MacGillivray *et al.* (1970)
		Hypopituitarism	Plasma	—	157	MacGillivray *et al.* (1970)
		Hypothyroidism	Plasma	189	112	Taylor *et al.* (1969)
		Hyperthyroidism	Plasma	346	212	Taylor *et al.* (1969)
Oxytocin	F and M		Plasma	1940	—	Fabian *et al.* (1968)
Insulin	F and M		Plasma	165	—	Silvers *et al.* (1969)

[a] Compilation of values published before 1968.

pound. This can be done by simultaneous intravenous injection of tritiated hormone dissolved in plasma and [^{14}C]-labeled hormone in saline solution followed by measurement of the [^{3}H] : [^{14}C] ratios of the hormone isolated from plasma samples at various intervals after the injection.

Because the concentration of the labeled hormone (c^*) can vary within the vascular system, the value of MCR measured by isotopic methods is dependent in principle upon the site of sampling. This situation does not compromise the usefulness of the parameter if c^* and c are measured in samples drawn from the same or equivalent sites since $PR = MCR \times c = I^* c/c^*$. It is then important to specify the site from which blood used to estimate MCR's was drawn, since arterial and peripheral venous blood may not necessarily have identical hormonal levels.

The metabolic clearance rate of a hormone is usually described as the volume of blood irreversibly cleared of the hormone per unit of time. However, it is possible to find situations in which MCR is larger than the cardiac output, as noted during studies of fetal metabolism of progesterone in the sheep (GURPIDE *et al.*, 1972). This apparent difficulty in interpreting values of MCR is minimized by noting the operational definition of MCR as the ratio of the production rate (rate of the *de novo* release of the hormone into circulation) to the concentration of the hormone.

Obviously, not only the site of sampling but also the site of injection of the tracer is important since it is fundamental that tracer and tracee have identical metabolic fates. As already discussed, peripheral intravenous injections are adequate for most but not all hormones.

Sources of a Circulating Hormone

The sources of a circulating compound can be investigated in one-tracer experiments by measuring "contribution factors" (Δ's), defined in Chapter 1 (Eqs. 1.15, 1.28, and 1.36). This parameter is obtained by comparing the specific activities of "precursor" and "product" in circulation (Eq. 2.18, Table 3.1). The examples given in Chapter 1 (Figs. 1.5 and 1.7; Table 1.2) serve to illustrate the method for estimation of conversion factors. The specific activities of estrone and estradiol in plasma, determined by LONGCOPE *et al.* (1968) during i. v. infusion of [^{3}H]estradiol in young men, indicate that only 14% of the circulating estrone derives from estradiol ($\Delta_{E_2E_1} = 0.14$). In order to further elucidate the source of estrone in blood, [^{3}H]androstenedione has been administered and the specific activities of circulating androstenedione were compared with the specific activities of estrone in blood (LONGCOPE *et al.*, 1969), or of estrone glucuronide in urine (MACDONALD *et al.*, 1971). It was found that a large part of the estrogen in men derives from the circulating androgen.

Measurement of Blood Flow

The basic formulas used for the measurement of blood flow by tracer methods can be easily derived. Consider that at site 1 in the vascular system, blood flows at a rate Φ_1 (ml/min) and a tracer is infused at a rate I_1^* (dpm/min). Then, the concentration of the tracer at a downstream site (Site 2) is

$$c_2^* = \frac{I_1^*}{\Phi_1}$$

if all blood passing through 2 originates from 1, if no removal or addition of the labeled compound in blood occurs between 1 and 2, and if there is no recycle of the tracer between the two sites. Obviously, c_2^* will remain the same whether $\Phi_2 \leqq \Phi_1$; therefore, not all the blood passing through site 1 is required to reach site 2.

From the equivalence between constant infusion and single injection modes of tracer administration expressed by Eq. 3.7, it follows that

$$\Phi_1 = \frac{I_1^*}{c_2^*} = \frac{S_1^*}{\int\limits_0^\infty c_2^*(t)\,dt}$$

where S_1^* is the dose injected into site 1, and $c_2^*(t)$ is the concentration measured at various intervals at site 2.

The condition of no recycle of the tracer may be physically fulfilled when a rapidly diffusible compound, e.g., an inert radioactive gas is used (LASSEN and LARSEN, 1972). Even when the tracer stays in circulation, concentrations determined before recycling can be used to estimate $\int_0^\infty c^*\,dt$ by extrapolation to infinity (KINSMAN *et al.*, 1929). More formal treatments to interpret data which includes recycling of the tracer have been described (BASSINGTHWAIGHTE, 1967, 1970; LARSON and SNYDER, 1972) and are based on principles of renewal theory (COX and MILLER, 1965). An experimental approach to the problem of recirculation of the tracer involves administration of different isotopes at sites 1 and 2 (ZIERLER, 1963).

A method to estimate the blood flow through an endocrine gland is based on the measurement of the production rate of a circulating hormone which is known to originate solely from that gland, and the determination of the arterial-venous differences in the concentration of the hormone across the gland. The blood flow Φ can be calculated from these data, since

$$PR = \Phi\,(c_v - c_a)\,.$$

This approach has been used by BAIRD to estimate the blood flow through the "active" ovary in women using estradiol as the secreted hormone (BAIRD, 1970).

It is also of interest to mention the "radioactive microsphere" method to measure the relative blood flow through various organs (RUDOLPH and HEYMANN, 1971). When microspheres of 15–50 μm are labeled with a radionuclide and introduced into circulation, the amounts of radioactivity trapped in different capillary beds reflect the distribution of blood flow to and within any organ.

B. Two Tracers Administered

Conversion Factors (ϱ's)

If a mixture of two metabolically related compounds is administered into a peripheral vein, either by infusion or as a single injection, the conversion factor (ϱ) of one to the other can be measured (Eqs. 2.14 and 2.15 for constant infusion experiments or Eq. 3.5 for experiments involving injection of the tracers). Note that the estimation of ϱ requires the experimental determination of the isotope ratio in only one of the compounds isolated from the blood samples. Since isotope ratios rather than concentrations are used, an arbitrary amount of the unlabeled compound to be isolated can be added to the samples to facilitate the purification procedures.

Suppose that a mixture of tritiated compound 1 and [^{14}C]-labeled compound 2 is introduced intravenously for the purpose of estimating the extent of conversion of 1 to 2. The isotope ratio in blood-borne compound 2 is the result of the overall conversion of 1 to 2 and is unaffected by the rate of metabolism of 2. The [^{14}C] tracer serves in this case as an "*in vitro* internal standard" to correct for metabolic losses of 2, as was lucidly expressed by MacDonald *et al.* (1967).

Some investigators choose to infuse one tracer at a time, measure the resulting steady-state concentrations of the labeled compounds in plasma and determine a "conversion ratio", which equals the quotient of the concentrations of isotope in product and precursor. Infusion of a tracer of the second compound yields a value of the conversion ratio in the opposite direction. From these data the ϱ's can be calculated (Eqs. 2.14 and 2.2) since

$$\varrho_{12} = \frac{I_2^*}{I_1^*}\,\frac{c_2^1}{c_2^2} = \frac{\mathrm{MCR}_2\, c_2^2\, c_2^1}{\mathrm{MCR}_1\, c_1^1\, c_2^2}$$

$$= \frac{\mathrm{MCR}_2}{\mathrm{MCR}_1}\,\frac{c_2^1}{c_1^1} = \frac{\mathrm{MCR}_2}{\mathrm{MCR}_1} \times \text{conversion ratio of 1 to 2}$$

and

$$\varrho_{21} = \frac{\mathrm{MCR}_1}{\mathrm{MCR}_2}\,\frac{c_1^2}{c_2^2} = \frac{\mathrm{MCR}_1}{\mathrm{MCR}_2} \times \text{conversion ratio of 2 to 1}\,.$$

The measurement of concentrations of the tritiated compounds in single tracer experiments usually involves addition of [^{14}C]-labeled indicators to the blood samples to estimate losses during the purification procedures. When the experimental design chosen calls for the administration of a mixture of two tracers differently labeled, carriers (unlabeled compounds) added to the blood samples serve the same purpose.

The single tracer experiments, which yield estimates of the MCR of the tracee and conversion ratios to other compounds, succeed in supplying average values for normal subjects. From these data and using average concentration values of the endogenous hormone, mean PR's and ϱ's in normal subjects are estimated. The two tracer experiments, on the other hand, obviously yield more accurate values for any particular subject (see Note 1.1).

Urinary data may be used to estimate transfer factors of compounds in circulation (Chapter 6). For instance, intravenous injection of a mixture of [^{14}C]dehydroisoandrosterone sulfate and [^{3}H]-16α-hydroxydehydroisoandrosterone-3-sulfate and measurement of the isotope ratio in the latter compound isolated from urine collected for several days following the injection, allowed the estimation of the transfer factor between the circulating compounds (GURPIDE *et al.*, 1973). The urinary product, in this case identical to the product in circulation, provides a "physical integration" of the isotope ratios of the compound in blood (TAIT, 1963). In a similar experiment, in which labeled dehydroisoandrosterone and dehydroisoandrosterone sulfate were injected intravenously, the ϱ of conversion of the unconjugated to the sulfate steroid in circulation was determined by analysis of blood and urine samples and the values obtained were found to be in reasonable agreement (SANDBERG *et al.*, 1964).

When the urinary metabolite is not identical to the circulating compound under study the situation is different. Suppose that 1 and 2 are the two blood-borne compounds under investigation. Compound 2 formed in a particular tissue from compound 1 may be released only partially to circulation since it may undergo immediate metabolism in the tissue. In this case, which is further discussed in Chapter 6, the transfer factor corresponding to the conversion of 1 to 2 in tissue may be greater than the overall transfer factor of 1 to 2 in circulation. This situation is evident when a mixture of labeled estradiol and estrone is administered and the isotope ratios of circulating estrone and of estrone glucuronide isolated from urine are compared. A much larger conversion of estradiol to estrone is detected from the urinary metabolite. This conversion corresponds to the oxidation of estradiol to estrone in tissue followed by conjugation of estrone in the same tissue, before passing to the bloodstream.

As described in the next section, transfer factors between circulating compounds are important for the estimation of the rates at which the hormones are produced from different sources, including glandular secretion. Some of the reported values of transfer factors between circulating hormones are listed in Table 5.2 (page 118).

Rates of Secretion and Metabolism

If the production rates (PR's) of the two circulating compounds under study and the conversion factors (ϱ's) between them are known, all the v's in Fig. 1.2, in which pools 1 and 2 may represent the compounds in systemic

Table 5.2. Transfer factors (ϱ's) between circulating hormones in normal subjects

From	To	ϱ (%)	Reference
Testosterone	Androstenedione	6.6 3.5–12 6.2 (M)–7.9 (F) 8.2 (M)–12 (F)	HORTON and TAIT (1966) RIVAROLA *et al.* (1967 a) MIGEON *et al.* (1968) LONGCOPE *et al.* (1969)
	Estrone	0.13	LONGCOPE *et al.* (1969)
	Estradiol	0.39 (M)–0.15 (F)	LONGCOPE *et al.* (1969)
	5α-Dihydrotesto-sterone	4.7 (M) 3.9 (M)–1.7 (F) 2.8 (M)–1.6 (F)	MAHOUDEAU *et al.* (1971) ITO and HORTON (1971) SAEZ *et al.* (1972)
Androstenedione	Testosterone	5.9 5–7 5.0 (M)–4.0 (F) 4.5 (M)–2.2 (F)	HORTON and TAIT (1966) RIVAROLA *et al.* (1967 b) MIGEON *et al.* (1968) LONGCOPE *et al.* (1969)
	Estrone	1.35 (M)–0.74 (F)	LONGCOPE *et al.* (1969)
	Estradiol	0.07	LONGCOPE *et al.* (1969)
	5α-Dihydrotesto-sterone	2.6 (M) 0.2–0.4	MAHOUDEAU *et al.* (1971) SAEZ *et al.* (1972)
Estradiol	Estrone	21	LONGCOPE and TAIT (1971)
	Estradiol-3-sulfate	42	LONGCOPE (1972)
	Estrone sulfate	65	RUDER *et al.* (1972)
Estrone	Estradiol Estrone sulfate	4.1–5 41 54	LONGCOPE *et al.* (1968) LONGCOPE (1972) RUDER *et al.* (1972)
Estrone sulfate	Estrone	21 15	RUDER *et al.* (1972) LONGCOPE (1972)
	Estradiol	2.2–4.4 1.4	LONGCOPE (1972) RUDER *et al.* (1972)
Dehydroisoandro-sterone	Dihydroisoandro-sterone sulfate	48 47 (31–59) 44 (32–51)	SANDBERG *et al.* (1964) MACDONALD *et al.* (1965) CHAPDELAINE *et al.* (1965)
	Androstenedione Testosterone	6.2 0.7	HORTON and TAIT (1967) HORTON and TAIT (1967)
Cortisol	Cortisone	58–62	DAZORD *et al.* (1971)
Cortisone	Cortisol	84–86	DAZORD *et al.* (1971)
Pregnenolone	Progesterone	2.4 7.3—26 (pregnancy)	LITTLE and BILLIAR (1969) LITTLE *et al.* (1971)
Progesterone	20α-Dihydro-progesterone	9.2	LIN *et al.* (1973)
Thyroxine	Triiodothyronine	46 (31–51)	SURKS *et al.* (1973)

circulation, can be estimated using the formulas shown in Table 1.1. At this point, we can recall the comments made in Chapter 1, viz.

1. The rates of exchange v_{12} and v_{21} in Fig. 1.2 include all paths by which circulating compounds 1 and 2 are interconverted, regardless of the sites where such conversions occur.

2. v_{10} and v_{20} correspond to the rates of irreversible removal of these compounds from circulation.

3. The rate at which a compound (e.g., compound 1) enters *de novo* the circulation (PR_1) can be resolved into the rate at which it is formed from the other circulating compound ($\varrho_{21}\, v_{02}$) or from all other sources (v_{01}). The latter may or may not correspond to the rate of secretion of the hormone by endocrine glands. It is always possible that another secreted precursor is converted to the hormone by extraglandular metabolism. The true secretion rate of the hormone can be estimated by a process of elimination using tracers of all probable secreted precursors.

Studies on the rates of interconversion and metabolism of blood-borne testosterone and androstenedione can be used as examples to illustrate these calculations. When Rivarola *et al.* (1966) administered intravenously tritiated testosterone (T) and [^{14}C]androstenedione (A) at a constant rate to a young woman, they could estimate from steady-state data that

$$PR_T = \text{Infusion rate } [^3H]T/\text{sp. act. plasma } [^3H]T = 0.37 \text{ mg/day}$$
$$PR_A = \text{Infusion rate } [^{14}C]A/\text{sp. act. plasma } [^{14}C]A = 2.7 \text{ mg/day}$$

and that 4.5% of the infused T is converted to A

$$(\varrho_{TA} = [^3H] : [^{14}C]_A/[^3H] : [^{14}C] \text{ infused} = 0.045)$$

and 5.2% of infused A is converted to T

$$(\varrho_{AT} = [^3H] : [^{14}C] \text{ infused}/[^3H] : [^{14}C]_T = 0.052)\,.$$

From these results, it can be calculated (Table 1.1) that 0.017 mg/day of circulating T is converted to circulating A, that 0.14 mg/day of A is converted to T, that 0.23 mg/day of T not derived from A enters the circulation *de novo*, presumably by secretion, and that 2.49 mg/day of A not derived from circulating T is released into the bloodstream. Therefore, whereas 38% of circulating T originates from circulating A, only 0.6% of A derives from blood-borne T. Similar experiments were performed by Horton and Tait (1966) and by Bardin and Lipsett (1967). Another example of this type of calculation was given in Chapter 1 (Fig. 1.5) using data obtained by Longcope *et al.* (1968) who administered labeled estrone and estradiol.

Pathways of Conversion of a Circulating Hormone to Another

Information about the pathways of conversion of a circulating compound to another can also be obtained from two-tracer experiments. Thus, if a tracer of compound 1 is administered intravenously, the contribution of 1 to a circulating compound 3 (Δ_{13}) can be estimated by comparing the specific activities of 1 and 3. When a tracer of an intermediate in this conversion (compound 2) and a tracer of compound 1 are administered simultaneously, and the specific activities of compounds 1, 2, and 3 are compared, the extent of the conversion of 1 to 3 through circulating compound 2 can be determined by using Eq. 2.21 or its equivalent for rapid injection of the tracers.

Fetomaternal Transfer and Production of Hormones

Pregnant animals with catheters inserted into fetal and maternal vessels can be used in experiments in which the same hormone labeled with different isotopes is administered simultaneously to mother and fetus. The isotopic data obtained from blood sampled through other catheters can then be applied to the estimation of rates of fetomaternal transfer and production of hormones. For these calculations it is sufficient to consider that in Fig. 1.2 and in Tables 2.1 and 3.1, pool 1 represents the blood-borne hormone in the fetus, whereas pool 2 represents the hormone in maternal circulation. Examples of applications of such an experimental design can be found in the literature (DIXON *et al.*, 1970; BEITINS *et al.*, 1970; GURPIDE *et al.*, 1972). An analysis of the sensitivity of the calculated parameters to changes in the data within the experimental error range has been reported by DI STEFANO *et al.* (1973) who used a computer-implemented Monte Carlo technique.

It is clear that in studies of a hormone delivered to the fetus only through the placenta (e.g., progesterone), administration of a tracer into an umbilical vein results in adequate labeling of the tracer. In contrast, a hormone such as cortisol, which is produced by the fetal adrenals as well as supplied to the fetal circulation through the placenta from maternal sources, adequate labeling of the blood-borne hormone in the fetus is difficult because a tracer infused through the umbilical vein may not have the same fate as the product of adrenal secretion.

It is important to ascertain that the fetal condition remains unchanged during the experiment. Fetal stress results in changes in patterns of blood distribution and alters quantitatively the metabolism of compounds (GURPIDE *et al.*, 1972).

With these limitations in mind, data on steady-state specific activities or areas under specific activity curves can be used to determine various rates of transfer, metabolism, and production of hormones in mother and fetus, according to Fig. 1.2. In this figure, the rates v_{12} and v_{21} correspond to rates of transfer of the hormone from one circulation to the other, by all paths, including passages through the placenta and fetal membranes not only of the hormone itself but also of compounds to which it may be reversibly converted. For instance, estradiol delivered from the placenta to the fetal circulation may find its way to maternal circulation through the following series of transformations: oxidation to estrone and sulfation of estrone by the fetus, hydrolysis of estrone sulfate by the placenta, and reduction of estrone to estradiol by the mother.

The rate v_{01} corresponds to the rate at which the hormone enters *de novo* the fetal circulation from sources other than the maternal circulation. For compounds such as progesterone, this rate may be clearly identified with the rate of secretion towards the fetus of the product of placental synthesis. Similarly, v_{02} may correspond to the rate of direct delivery to the maternal circulation of the hormone synthesized in the placenta if all other sources of maternal production of the hormone can be excluded. By measuring v_{01} and v_{02}, it could be possible to estimate the relative rates at which a hormone made in the trophoblast is released toward the maternal and fetal sides, an important and still unclear parameter of hormonal dynamics.

Rates of Metabolism in Specific Organs

Two different experimental designs to estimate rates of metabolism of the hormone in a particular organ are now discussed. In one of these designs, a labeled hormone is administered directly into the arterial supply of the organ while a differently labeled tracer of the same hormone is given intravenously. In the second method, a mixture of two metabolically related compounds, labeled with different isotopes, is administered intravenously.

Method 1. Consider the case in which a hormone labeled with one isotope is infused into the arterial supply of an organ and, simultaneously, the same hormone labeled with another isotope is infused into a peripheral vein. This type of experiment can be analyzed on the basis of Fig. 1.2, where pool 1 now represents the compound in the organ into which tritiated tracer is infused, and pool 2 represents the compound in peripheral circulation. The extraction of the hormone by the organ can be calculated from the isotope ratios in the infused tracer mixture and in the hormone isolated from peripheral blood after a steady state is reached. The isotope ratio in the blood-borne compound reflects the irreversible removal of the tracer by one passage through the organ into which it was directly infused. Furthermore, as indicated in Table 1.1, the rate of irreversible metabolism of the hormone in the organ (v_{10}) or elsewhere (v_{20}), and the rates of de novo formation of the hormone in the organ under study (v_{01}) or elsewhere (v_{02}) could be calculated from specific activity data on the compound in the organ and in peripheral circulation. Although measurements of the specific activity of the hormone in tissue appears unfeasible, it is possible to measure the specific activity of a circulating metabolite formed exclusively from the hormone in that organ. Such a metabolite would have the same steady-state specific activity as its precursor.

The origin, i.e., organ-versus-extraorgan formation, of any other circulating metabolites could be elucidated according to the calculations described by Eqs. 2.21 and 2.23.

An application of this experimental design can be illustrated by referring again to studies involving infusions of labeled progesterone into sheep fetuses *in utero* (GURPIDE *et al.*, 1972). In these studies, a fetus was partially delivered through a uterine incision and catheters were introduced into a superior mesenteric vein, a femoral vein, and a femoral artery. Tritiated progesterone was then infused directly into the liver through the mesenteric catheter, while [^{14}C]-labeled tracer of the hormone was infused into the femoral vein. The isotope ratio in progesterone isolated from blood drawn from the femoral artery indicated a 70% hepatic extraction of the hormone. A similar experiment in which the tritiated progesterone was infused into the umbilical vein rather than directly into the liver gave results indicating that in one passage about 70% of the placental hormone reaches the inferior vena cava, the site where [^{14}C]progesterone was introduced. By comparing these results with those of hepatic extraction, it was estimated that approximately 60% of the placental flow was channeled through the ductus venosus, while 40% passed through the liver in the animals studied.

A basically identical experimental design is used in studies in which a labeled (e.g., [^{14}C]) compound is administered orally, and at the same time another tracer (e.g., [^{3}H]-labeled) of the same compound is injected intravenously. Comparison of the [^{3}H] and [^{14}C] specific activity areas of the compound isolated from peripheral blood or urinary metabolites indicates the fraction of the ingested tracer transferred to the bloodstream (HORTON and TAIT, 1966; TAIT and LITTLE, 1968). For example, after administration of labeled dehydroisoandrosterone sulfate ([^{14}C] orally and [^{3}H] intravenously) to pregnant women, isotope ratios were measured in urinary dehydroisoandrosterone sulfate and in urinary estrogens (GURPIDE and VANDE WIELE, 1971). As expected, the isotope ratios in these compounds were identical since the labeled estrogens were formed in the placenta from circulating dehydroisoandrosterone sulfate. Note, however, that a metabolite of splanchnic origin could have a lower isotope ratio than the hormone in circulation.

Method 2. In another experimental design, the labeled hormone and a tracer of a metabolically related compound, labeled with another isotope, are mixed and the mixture is infused intravenously. The steady-state concentrations of both labeled compounds are measured in samples of peripheral blood and the venous effluent of the organ under consideration. Comparison of concentrations and isotope ratios in these two samples allows the estimation of the fraction of each hormone entering the organ that is either irreversibly removed through one passage or is converted to the other compound.

The derivation of the formulas used in these calculations can be easily followed with the help of the scheme in Fig. 5.1, which illustrates the renal metabolism of testosterone (CHAPDELAINE, 1969). Tritiated testosterone (T) and [^{14}C]androstenedione (A) are infused intravenously, and concentrations of labeled T and A are measured in peripheral blood and in the renal venous effluent, at the isotopic steady state.

The transfer factor ϱ_{av}^{TT} characterizing the passage of T through the organ is given by the expression

$$\varrho_{av}^{TT} = \frac{c_{Tv}^{^3H} - \varrho_{av}^{AT} \times c_{Aa}^{^3H}}{c_{Tv}^{^3H}}. \tag{5.1}$$

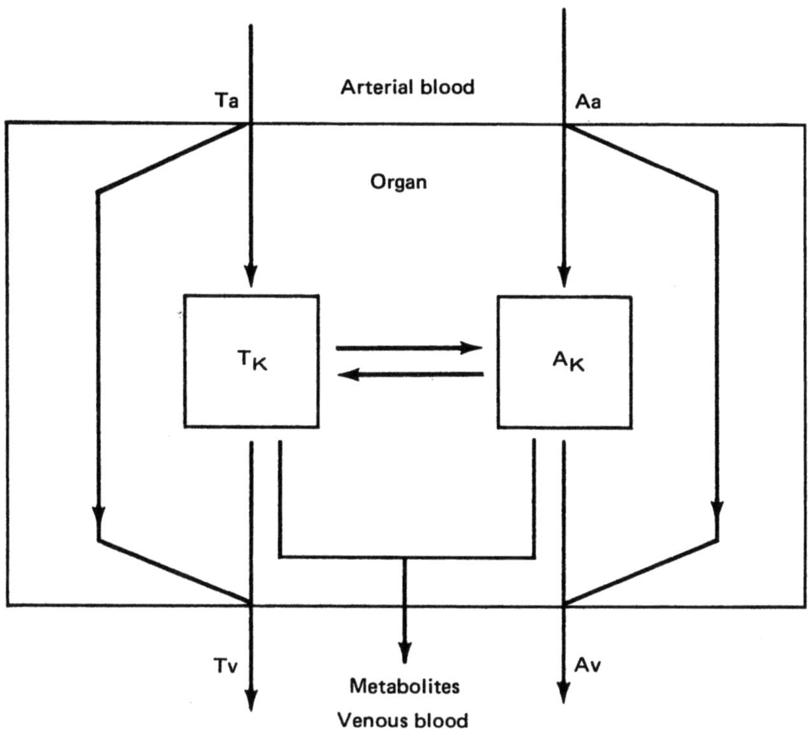

Fig. 5.1. Scheme illustrating processes involved in the transfer of two interconvertible hormones (testosterone T, and androstenedione A) across an organ

The subindices a and v denote artery and vein, respectively. Note that the numerator in this expression indicates the concentration in the venous effluent of ^3H-T derived from arterial ^3H-T. This value is obtained by subtracting from the total concentration of ^3H-T in the venous sample the amount contributed by ^3H-A converted to ^3H-T in the kidney.

The same relation applies to the other isotope:

$$\varrho_{av}^{TT} = \frac{c_{Tv}^{^{14}C} - \varrho_{av}^{AT} c_{Aa}^{^{14}C}}{c_{Ta}^{^{14}C}}.$$

The last two equations can be combined to determine the value of the factor ϱ_{av}^{AT} corresponding to the conversion of arterial A to venous T. Thus,

$$\varrho_{av}^{AT} = \frac{c_{Tv}^{^3H} c_{Ta}^{^{14}C} - c_{Tv}^{^{14}C} c_{Ta}^{^3H}}{c_{Aa}^{^3H} c_{Ta}^{^{14}C} - c_{Aa}^{^{14}C} c_{Ta}^{^3H}}. \tag{5.2}$$

By multiplying this equation by $c_{Ta}^{14C}\,c_{Tv}^{14C}/(c_{Ta}^{14C}\,c_{Tv}^{14C})=1$

$$\left(\varrho_{Av}^{AT}=c_{Tv}^{14C}\,\frac{(^3H/^{14}C)_{Tv}-(^3H/^{14}C)_{Ta}}{c_{Aa}^{3H}-(^3H/^{14}C)_{Ta}\,c_{Aa}^{14C}}\right)$$

and then by $c_{Aa}^{14C}/c_{Aa}^{14C}=1$, it follows that

$$\varrho_{av}^{AT}=\frac{c_{Tv}^{14C}}{c_{Aa}^{14C}}\left[\frac{(^3H/^{14}C)_{Tv}-(^3H/^{14}C)_{Ta}}{(^3H/^{14}C)_{Aa}-(^3H/^{14}C)_{Ta}}\right]. \tag{5.3}$$

An equation for ϱ_{av}^{Ta} (Ta to Av in Fig. 5.1) can be similarly derived to yield

$$\varrho_{av}^{TA}=\frac{c_{Av}^{14C}}{c_{Ta}^{14C}}\left[\frac{(^3H/^{14}C)_{Av}-(^3H/^{14}C)_{Aa}}{(^3H/^{14}C)_{Ta}-(^3H/^{14}C)_{Aa}}\right]. \tag{5.4}$$

The transfer factor ϱ_{av}^{TT} can be calculated by substituting ϱ_{av}^{AT} (Eq. 5.3) in Eq. 5.1. Similarly, the value of ϱ_{av}^{TA} estimated from Eq. 5.4 can be used to calculate ϱ_{av}^{AA} (A_a to T_v in Fig. 5.1):

$$\varrho_{av}^{AA}=\frac{c_{Av}^{3H}-\varrho_{av}^{TA}\,c_{Ta}^{3H}}{c_{Aa}^{3H}}.$$

If the same fractions of T and A brought into the kidneys by the arterial supply enter the cells, then the intracellular conversion factors ϱ_{KK}^{AT} and ϱ_{KK}^{AT} (T_K to A_K in Fig. 5.1) can also be estimated from data on concentrations of the labeled compounds in arterial and venous blood at the isotopic steady state. Thus,

$$(^3H/^{14}C)_{TK}=\frac{c_{Ta}^{3H}+\varrho_{KK}^{AT}\,c_{Aa}^{3H}}{c_{Ta}^{14C}+\varrho_{KK}^{AT}\,c_{Aa}^{14C}} \tag{5.5}$$

or

$$\varrho_{KK}^{AT}=\frac{c_{Ta}^{3H}-(^3H/^{14}C)_{TK}\,c_{Ta}^{14C}}{(^3H/^{14}C)_{TK}\,c_{Aa}^{14C}-c_{Aa}^{3H}}. \tag{5.6}$$

Similarly,

$$\varrho_{KK}^{TA}=\frac{c_{Aa}^{3H}-(^3H/^{14}C)_{AK}\,c_{Aa}^{14C}}{(^3H/^{14}C)_{AK}\,c_{Ta}^{14C}-c_{Ta}^{3H}}.$$

This type of calculation has been described and applied to the study of the metabolism of steroids in various organs by CHAPDELAINE (1969) and RIVAROLA *et al.* (1967 a).

C. Three Tracers Administered

The application of two-tracer methods to estimate rates of interconversion between circulating compounds or rates of transfer of hormones between fetal and maternal circulations has been described in the preceding section. In order to obtain quantitative information about the paths by which such conversions or transfers occur, a third tracer may be used.

Consider the scheme in Fig. 1.6. Pools 1, 2, and 3 may correspond to three different compounds in circulation. Formulas to be used for the calculation of rates of formation, interconversion and metabolism of these compounds are given in Table 1.3.

Note that the use of three tracers yields further information about the origin of the compounds involved. When only labeled compound 1 is administered, all that can be found is the total rate at which it appears *de novo* in circulation (PR_1, Eqs. 2.1 or 3.2), regardless of its origin. If tracers of compounds 1 and 2 are used, it is possible to determine which part of PR_1 is derived from 2 ($\Delta_{21}^{(2)}$) and therefore from sources other than circulating compound 2 ($v_{01}^{(2)}$, Tables 2.1 and 3.1). When labeled compounds 1, 2, and 3 are administered, it is possible to estimate the rate of *de novo* entry of 1 into circulation from sources other than 2 and 3 ($v_{01}^{(3)}$). This rate is obviously closer to the true rate of glandular secretion of the hormone than are either PR_1 or $v_{01}^{(2)}$. The formulas to calculate $v_{01}^{(3)}$ or other rates pertaining to a three-pool system are presented in Table 1.3.

Figure 1.7 in Chapter 1 shows an example based on data taken from the literature (LONGCOPE, 1972). The three metabolically related blood-borne compounds in that example are estrone, estradiol, and estrone sulfate. The values for production rate and conversion factors were obtained from constant infusion of tracers (estrone and estradiol) or by single-injection experiments (estrone sulfate).

II. Data: Specific Activity Functions

In addition to the parameters described in the preceding section (PR's, MCR's, ϱ's, Δ's, v's) various others can be calculated when specific activity functions of blood-borne compounds are determined. These parameters, which further characterize the distribution and metabolism of a hormone, have been discussed in Chapter 4.

Methods for determination of tracer concentration or specific activity functions of hormones in blood were described in Chapter 3. The graphic "peel-off" procedure and the computer-implemented numerical methods discussed in that chapter presupposed that the functions are sums of exponential terms with real exponents ($\sigma(t) = \Sigma D_i e^{-\alpha_i t}$). Although adequate fitting to the experimental data is obtained by these procedures, it should be kept in mind that other types of functions may also be compatible with the data, e.g., $\sigma = \Sigma A_i t^{B_i} e^{-\alpha_i t}$ (HOWARD *et al.*, 1969), or $\sigma = \Sigma A_i t^{B_i}$ (ANDERSON *et al.*, 1969).

Note 5.1 illustrates the calculation of specific activity functions from experimental data. For this purpose, values of concentrations of tritiated estradiol in plasma resulting from the injection of a tracer of this hormone to nonpregnant women (LONGCOPE and WILLIAMS, 1974), were used. We are grateful to Dr. LONGCOPE for making these results available to us before publication. The analysis of these data is particularly pertinent because it complements other illustrative examples involving the estrogen system which are given throughout the monograph. The reader may evaluate from the example in Note 5.1 the degree of accuracy with which specific activity functions, and the parameters calculated from them, can be determined.

The following parameters can be estimated when a tracer is injected intravenously and its specific activity is measured in blood samples taken at various intervals after infusion.

Size of the Pool of Fast Initial Distribution of the Intravenously Injected Tracer

The amount of material (M) in the pool into which the intravenously injected tracer is distributed can be estimated from the value of σ at $t=0$ (σ_0), which equals the ratio Dose/M. Then, $M = \text{Dose}/\sigma_0$. As shown by Eq. 3.16, if $\sigma(t)$ is a sum of exponential terms, $\sigma_0 = \Sigma D_i$. However, uncertainties involved in the extrapolation of the specific activity function to $t=0$ are expected as time must be allowed for distribution of the radioactivity injected as a bolus into the vascular system. This problem is more clearly discussed by considering another parameter commonly used in *in vivo* tracer experiments: the so-called volume of the space of fast initial mixing of the tracer.

"Volume" of the Space of Fast Initial Distribution

This volume can be formally defined as the ratio of the size of the fast mixing pool (M) to the peripheral concentration of the circulating compound (c). Thus, $V = M/c$.

If c is the concentration of the hormone in peripheral plasma, V represents a volume in which a hormone uniformly distributed at that concentration will contain the amount of material M. The value of V is usually determined by plotting plasma concentration values of the labeled hormone (c^* in dpm/ml) against the time elapsed after the injection, extrapolating the resulting curve to $t=0$. In this case, $V = \text{Dose}/c_0^*$. This value is often considerably larger than the plasma volume, even when sampling is begun within a few minutes following the injection of the tracers. In these cases, it can be concluded that the compound leaves the vascular system rapidly. However, the calculated values of V cannot be considered to be accurate since they are very sensitive to the timing of the early samples and to the precision of the data (see Note 5.1). Since the concentrations of the tracer in plasma and in extracellular spaces are not necessarily equal, the apparent volumes of distribution of a hormone determined by tracer methods can be either larger or smaller than the extracellular water space.

Rates of Exit of a Compound from Circulation

The rate constant of exit of a hormone from the circulation (or from the space of fast initial mixing, which includes the vascular system) equals the ratio of the rate of exit ($v_{T,1}$) and the pool size (M_1), i.e., Eq. 3.22:

$$k_{T,1} = \frac{v_{T,1}}{M_1} = \frac{\Sigma D_i \alpha_i}{\Sigma D_i}.$$

Therefore, $k_{T,1}$ as well as $v_{T,1}$ can be estimated from the specific activity function (or the concentration function) of the labeled hormone in blood.

For compounds with a volume of fast initial distribution equal to the plasma volume, $v_{T,1}$ then represents the rate of transcapillary passage.

As is evident from Fig. 1.1, w_1 (the rate at which the hormone reenters the circulation or the fast mixing pool) can be calculated if the function $\sigma = \Sigma D_i e^{-\alpha_i t}$ is known, since (Eqs. 3.23 and 3.2)

$$w_1 = v_{T,1} - v_{10} = S_1^* \frac{\Sigma D_i \alpha_i}{(\Sigma D_i)^2} - \frac{S_1^*}{\Sigma (D_i/\alpha_i)}.$$

When tracers of two metabolically related hormones are injected intravenously, the rates of exit of each hormone from circulation ($v_{T,1}$ and $v_{T,2}$) can be estimated. Therefore, since their rates of interconversion and of irreversible removal can also be calculated from the specific activity functions (Table 2.1), it follows that (Fig. 1.2)

$$w_1^{(2)} = v_{T,1} - (v_{10}^{(2)} + v_{12}^{(2)})$$

and

$$w_2^{(2)} = v_{T,2} - (v_{20}^{(2)} + v_{21}^{(2)}).$$

These calculations can be illustrated using Longcope's data obtained after intravenous administration of labeled estrone, estradiol and estrone sulfate (Notes 5.2 and 5.3).

Reentry into Circulation

Part of the total amount of the compound leaving the circulation at any time is lost irreversibly whereas the rest returns to it. The fraction lost irreversibly (Φ) can be calculated without the postulation of models (Eqs. 3.25 and 3.26).

Average Times of Transit, Residence, and Recycling of Hormones in Circulation

The mean transit time of a compound in circulation ($\bar{t}_1$) is defined as the average interval of time spent by a molecule of the compound from its entry into the vascular system to its next exit from it. Thus, if a molecule enters the bloodstream at $t = 0$, $\bar{t}_1$ is the expected time of its first exit. In other words, if all molecules were to spend the same time, $\bar{t}_1$, in each passage through circulation, the velocity at which they leave the vascular system ($v_{T,1}$) equals the ratio of the amount of the compound in circulation (M_1) to the time necessary to remove this amount of material ($\bar{t}_1$), i.e., $v_{T,1} = M_1/\bar{t}_1$ or $\bar{t}_1 = M_1/v_{T,1}$. The mean transit time of a hormone through circulation can be estimated if the curve of concentration of the labeled hormone in blood is determined, as indicated by Eq. 3.28. If the space of fast initial distribution of the hormone entering the circulation is larger than the plasma volume, the mean transit time corresponds to the stay of the hormone in this larger space. In this case, however, the calculations are less reliable due to the previously discussed inaccuracies which can be expected in the estimation of values of concentrations and derivatives of concentration functions at $t = 0$.

The mean total residence time of a hormone in circulation ($\bar{T}_1$) can be defined as the average total interval of time spent by a molecule of the compound in the vascular system (or in the space of fast initial distribution) in all passages through it. It can be considered that if all molecules behave identically, they would each stay in circulation for a total period of time $\bar{T}_1$ before leaving the system irreversibly. Accordingly, the ratio $M_1/\bar{T}_1$ represents the rate of irreversible removal of the compound from circulation. Note that this rate corresponds to the "production rate" of the hormone in circulation (Eq. 3.27).

The mean recycling time of a circulating hormone ($\bar{t}_{11}$) is defined as the average interval of time spent by a molecule of the compound from its exit from the vascular system (or the space of fast initial distribution) to its next reentry into it. A formula to calculate the mean recycling time of a hormone from isotopic data can be obtained by referring to Eq. 3.32.

Some of the values of the times of transit, residence, and recycling of circulating hormones, computed from published data, are shown in Table 5.3, which lists only the values of parameters corresponding to hormones for which the analysis of tracer concentration curves yields distribution volumes comparable to the plasma volumes.

The mean interval of time involved in a single, unidirectional conversion of one circulating hormone to another ($\bar{t}_{12}$ or $\bar{t}_{21}$) can be estimated from isotopic data when tracers of two hormones are administered intravenously, according to the formulas presented in Chapter 3 (Eqs. 3.39 and 3.40). In that chapter formulas are also given to estimate the mean time of recycle of a hormone by pathways that do not involve intermediacy of the other hormone in circulation (Eqs. 3.37 and 3.38). These calculations may provide new parameters to assess metabolic events under physiologic and pathologic conditions. It should be noted that measurement of transfer times can also be applied to the study of the movement of hormones between fetal and maternal circulations.

Table 5.3. Average times of transit ($\bar{t}_1$), recycle (t_{11}), total residence (T_1) and total recycle ($\overline{T}_{11}$) of some hormones in systemic circulation

Compound	Volume of distribution (liters of plasma)	$\bar{t}_1$ (h)	$\bar{t}_{11}$ (h)	$\bar{t}_1/\bar{t}_{11}$	$\overline{T}_1$ (h)	$\overline{T}_{11}$ (h)	φ_1	ν_1	References
Human chorionic gonadotropin	2.2	8.9	6.3	1.4	14	9.9	0.64	0.56	Rizkallah *et al.* (1969)
Human placental lactogen	3.3	0.34	0.34	1.0	0.63	0.64	0.53	0.88	Kaplan *et al.* (1968)
Thyrotropin	2.7	0.38	0.27	1.4	1.2	0.83	0.33	2.0	Beckers and Cornette (1971)
Insulin	2.8	0.08	0.38	0.21	0.41	2.0	0.19	4.2	Silvers *et al.* (1969)
Thyroxine	3.1	0.53	1.5	0.35	21	61	0.025	39	Cavalieri and Searle (1966)
Testosterone sulfate	2.5	0.40	0.76	0.53	0.74	1.4	0.54	0.85	Saez *et al.* (1971)
Dehydroisoandrosterone	1.7	0.29	0.50	0.58	2.1	3.6	0.14	6.2	Sandberg *et al.* (1964)
Cortisol	2.4	0.29	0.31	0.94	3.3	3.5	0.09	10	Wang *et al.* (1967 b)
Basal	2.8	0.13	0.30	0.43	0.58	1.30	0.23	3.3	Bradley and Waterhouse (1966)
Estrogen treatment	2.4	0.25	0.18	1.4	1.11	0.80	0.22	3.5	Bradley and Waterhouse (1966)

III. Calculations Based on Models Involving a Limited Number of Compartments

All the parameters discussed in the preceding section (M's, V's, k_T's, v_T's, w's, φ's, turnover times) may be estimated from the isotope content of compounds in circulation, making no assumptions about the number, identity or characteristics of pools with which they exchange isotope.

Another procedure currently used to obtain information from specific activity data of labeled compounds in blood is the following.

1. A function consisting of a sum of exponential terms fitting the experimental data is determined.

2. A model is built in which the number of interconnected pools equals the minimum number of exponential terms that describes the data adequately. Since usually the experimental curve is satisfactorily fitted by a function consisting of two to four exponential terms, two to four pool models are then considered.

3. Rate constants corresponding to various processes indicated in the model are estimated using the values of the coefficients and exponents (Chapter 4). Because often the available data is insufficient to calculate all of these rate constants, procedures have been developed (BERMAN and SCHOENFELD, 1956) to map solutions compatible with the physical constraints of the model, e.g., rate constants must not be negative or larger than the rate constants of total removal of the compound from blood.

In addition to the previously discussed experimental imprecisions, which lead to uncertainties in the number of exponential terms necessary to describe the data, the following two criticisms to this approach can be mentioned.

1. Even though the chosen model may be the simplest, an unlimited number of more complex models are equally compatible with the experimental data. Because some kind of anatomic or physiologic counterpart for a pool is usually proposed, it seems reasonable to question the soundness of choosing the simplest model to represent the physiologic situation.

2. From a formal point of view, a pool in the model indicates a space of fast mixing of a compound. However, no realistic counterpart for a model consisting of only few pools can be expected as no fast exchange occurs among cellular contents, even in a single organ. Since there are many different criteria to justify lumping individual fast mixing pools into entities that behave as a pool, the definition of the pool in a simple model is vague.

The disturbing result of this situation is that tracer experiments are forced to provide values for rates of transfer of the circulating compound to various peripheral pools even though the definition of either a peripheral pool or of a pool resulting from lumping of several others is not clear.

Relationships between rate constants in a system of n communicating pools, which result in only a few exponential terms in the specific activity function of the pool into which the tracer is injected is described in Chapter 4. These relations are now used in an attempt to clarify the interpretation of isotopic data from circulating compounds.

Equation 4.68, in which pool 1 represents the blood-borne hormone, may be used to answer the following question: can peripheral pools which have identical rate constants of irreversible removal (k_{i0}) and of transfer to circulation (k_{i1}) be combined insofar as they may contribute only one exponential term to the biexponential experimental curve? The answer to this question is affirmative; furthermore, it is found that no restrictions need be imposed upon the exchange between peripheral pools. Thus, if k_{i1} is constant for $i \geqq 2$ and k_{i0} is constant for $i \geqq 2$, Eq. 4.68 is satisfied. This is made evident by canceling k_{j1} and k_{i1} (i. $j \geqq 2$) in Eq. 4.68 and by using the relation $-k_{ii} = \Sigma^{n}_{\substack{j=0 \\ j \neq i}} k_{ij}$, i.e.,

$$\frac{\sum_{i=2}^{n} K_{1i}^{(q)} \sum_{j=2}^{n} k_{ij}}{\sum_{i=2}^{n} K_{1i}^{(q)}} = \sum_{j=2}^{n} k_{ij} = -(k_{i1} + k_{i0}) = \text{constant} \qquad i \geqq 2.$$

The constant in this equation represents the sum of the rate constants at which material is removed from any of the peripheral pools, by irreversible metabolism and by release into circulation ($k_{i0} + k_{i1}$).

If there is no exchange among the peripheral pools ($k_{ij} = 0$ for i, $j \geqq 2$, unless $i = j$), Eq. 4.68 becomes

$$\frac{\sum_{i=2}^{n} K_{1i}^{(q)} k_{ii}\, k_{i1}}{\sum_{i=2}^{n} K_{1i}^{(q)} k_{i1}} = k_{ii} = \text{constant} \qquad q = 1, 2, \ldots, n-1$$

and is satisfied if k_{ii} is the same for all $i \geqq 2$. The condition is less stringent than the one in the previous case since here it suffices that $k_{i1} + k_{i0}$ be constant, regardless of the individual values of k_{i1} and k_{i0}.

The hypothetical situation in which there is one organ, e.g., the liver, that accounts for most of the exchange of the hormone between blood and tissues can be used as an example. If there were no direct exchange of the hormone among the type of hepatic cells involved, the biexponential curve indicates that the rate at which it leaves each of the cells is proportional to the cellular content of the hormone, i.e., that the rate constant of removal is the same in each of these cells (the hormone in each cell is considered to be a different pool). If there were direct exchange of the hormone among the cells in question, it would be necessary not only that the sum of the rate constants of irreversible metabolism and of transfer back to circulation were identical for each cell but that the relative proportion between these two processes were also the same. In both cases, the value $[(D_1 \alpha_2 + D_2 \alpha_1) / (D_1 + D_2)]$, calculable from the experimental curve, measures the sum of the rate constants of irreversible removal and of transfer to circulation for each cell.

Furthermore, under the latter conditions ($k_{ij} = 0$, $k_{ii} = \text{constant}$, $i \geqq 2$), the specific activity functions of the hormone in the organ would also be

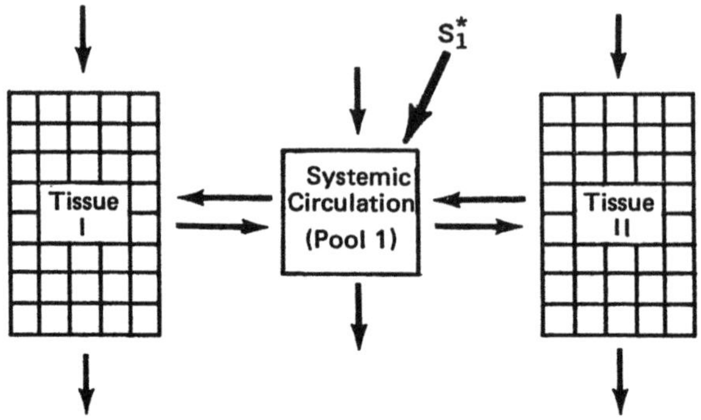

Fig. 5.2. A function $\sigma = D_{11} e^{-\alpha_1 t} + D_{12} e^{-\alpha_2 t} + D_{13} e^{-\alpha_3 t}$ describing the specific activity of a labeled hormone in circulation following its intravenous injection, may result from the following situation: all of the compound reentering the circulation originates from cells in tissue I and tissue II, the ratio between the specific activities in any two cells of tissue I or between any two cells of tissue II does not change with time. If these conditions prevail, the specific activities of the compound in tissues I and II are also given by a function of time consisting, at most, of a sum of three exponential terms with the same exponential constants (α_1, α_2, α_3). From GURPIDE and MANN (1970).

expected to be biexponential, with the same exponents, α_1 and α_2, as the function of the hormone is circulation. These statements follow from Eq. 4.77.

The conditions discussed in these examples are by no means necessary to satisfy Eqs. 4.68 or 4.77. Many other relationships among the k's would result in a curve of disappearance of tracer from circulation described by two exponential terms. This means that the use of a two-pool model to represent the system even with the recognition that the second pool may represent a lumping of pools having identical removal rates, is arbitrary since many other possible relations among the pools would be compatible with the experimental data.

The report from which Eq. 4.77 was taken (MANN and GURPIDE, 1969 a) also gives conditions that may be used to interpret specific activity functions with more than two exponential terms. These are necessary and sufficient conditions relating the rate constants in the strongly connected (irreducible) system which assure a linear dependence among the specific activities of the pools. If, for instance, the relations among the rate constants are such that the specific activity of each pool exchanging with the tracer in circulation can be expressed in terms of the specific activities of only two pools of the system, then each pool, including the circulating tracer, has isotope concentration curves described, at most, by three exponential terms. One prediction resulting from such an analysis is that any major organ exchanging the compound with blood will contribute one exponential term to the plasma specific activity curve, if the specific activities of the compound throughout each organ are either identical or proportional at all times (Fig. 5.2). Perhaps this is the case implicit in most interpretations of tracer data when anatomic counterparts are assigned to each of the few pools in the model. The present discussion suggests means to test for such an hypothesis by obtaining specific activity or isotope concentration curves simultaneously from various organs and different portions of the same tissue.

Furthermore, this analysis shows that pools belonging to different organs, and even physiologically unrelated regions, may have proportional specific activities as a result of peculiar values of their rate constants. These unrelated pools would, in fact, appear to be a single pool, as evaluated from the kinetics of disappearance of the tracer from plasma. Once again, the arbitrariness associated with the use of simple kinetic models is apparent.

In summary, the interpretation of isotope data from circulating hormones by means of models consisting of pools to which some physiologic meaning is assigned requires more formal justification than currently given. Analysis of relations among the rate constant corresponding to pools (intracellular hormones) exchanging isotope with the hormone in blood, allows the postulation of situations that would justify lumping pools and using relevant models. Furthermore, experimental tests to examine the validity of the postulated situations become self-evident when this approach is followed.

IV. Analysis of Systems that are Not at the Steady State

The rate of production of many hormones is not uniform and the assumption of the existence of an endogenous steady state is not always acceptable. In addition to the observed circadian periodicity in the secretion of some hormones, and hormonal changes during the estrous cycle, truly intermittent secretion of adrenal hormones has been reported (Hume, 1958; Berson and Yallow, 1968; Hellman *et al.*, 1970; Krieger *et al.*, 1971). Similar rapid fluctuations in the secretion of gonadotropins have been recently observed (Midgley and Jaffe, 1971; Yen *et al.*, 1972).

When the production of hormones is variable, two situations must be clearly distinguished from each other. In one of them, the rate constants corresponding to the various metabolic and transfer processes involving the hormone remain constant during the period of observation. Under these conditions, the formulas presented in the preceding chapters, properly modified to yield average rates, are applicable (see Note 3.2). In contrast, a situation in which rate constants, as well as rates, change with time is beyond the scope of this monograph. Such possibilities would require more elaborate experimental designs and mathematical techniques such as those employed in dynamic programming (BELLMAN, 1957).

The rate constant of a process equals, as defined before, the ratio of the corresponding rate to the amount of material in the pool from which the transfer originates. If rates and pool sizes change proportionally, the rate constants remain unchanged. It is apparent from the theory presented in Chapter 4 that the isotope content in each pool is given by rate constants rather than by rates. If y_2^1 denotes the total amount of the isotope introduced at $t=0$ into pool 1 which is present at time t in pool 2, it follows from Eq. 4.41 that

$$y_2^1 = \sigma_2^1 M_2 = S_1^* k_{12} \frac{e^{-\alpha_2 t} - e^{-\alpha_1 t}}{\alpha_1 - \alpha_2} \tag{5.7}$$

where α_1 and α_2 are determined by the rate constants of the system (Eqs. 4.28 and 4.29). Similarly, if x_2^1 is the amount of isotope achieved in pool 2 during infusion of the tracer into pool 1, it results from Eq. 4.60 that

$$x_{2\infty}^1 = a_{2\infty}^1 M_2 = \frac{I_1^* h_{12}}{M_1 \alpha_1 \alpha_2} M_2 = \frac{I_1^* k_{12}}{\alpha_1 \alpha_2}\,. \tag{5.8}$$

Consequently, parameters such as metabolic clearance rates (MCR, Eq. 2.2), transfer factors (ϱ, Eq. 2.14), fractional losses (φ, Eq. 3.26), turnover times (Table 3.2), and others that can be calculated from concentrations of the tracer in blood are independent of the rates of secretion of the hormone if the rate constants do not change and all processes are either first order or at a steady state. Note that the formulae used to calculate these parameters from isotopic data involve concentrations of labeled compounds in blood (dpm/ml) rather than specific activities. The constancy of the MCR's of hormones under physiologic conditions has been already discussed in this chapter.

Assuming that the rate constants of transfer and metabolism of a circulating hormone are concentration and time-independent under physiologic conditions, "instantaneous" production rates can be calculated from concentration data by inversion of the convolution integral

$$c(\tau) = \int_0^\tau f(\tau - t)\, PR(t)\, dt$$

in which $c(\tau)$ (μmole/liter) is the concentration of the hormone in blood at time τ, $f(\tau - t)$ (liter^{-1}) is the fraction of a unit dose of the hormone that entered the circulation at time t and which remains at τ in a liter of blood, and $PR(t)$ (μmole/min) is the production rate of the hormone at time t. The function $f(\tau - t)$ can be determined experimentally by injecting a single dose of the labeled hormone and measuring the concentration of the tracer in blood at various times. Deconvolution procedures can then be used to estimate the function $PR(t)$. REBAR *et al.* (1973) have used this approach to study luteinizing hormone secretion in humans.

TURNER *et al.* (1971) applied the convolution expression to the study of insulin delivery rates. They determined the function $f(\tau - t)$ by administering a dose of the hormone and by measuring the fraction of the peak concentration value remaining in plasma at various times after the injection. To estimate minute-to-minute delivery rates of insulin in systemic circulation, TURNER *et al.* took blood samples every minute ($\Delta t = 1$) and applied the discrete equation

$$c(\tau) = PR(\tau) + \sum_{t=1}^{\tau-1} f(\tau - t)\, PR(t)\,.$$

Other expressions to estimate the component of a circulating hormone derived from secretion occurring during the interval between two measurements, have been very simply obtained by assuming a single exponential decay. Thus, KRIEGER *et al.* (1971) calculated the "residual level" of plasma cortisol at a time τ using the equation.

$$c_{\tau,\,\text{residual}} = c_{(\tau-\Delta t)} - K\, c_{(\tau-\Delta t)}\, \Delta t$$

where Δt is the interval between τ and the time of the preceding observation and where K corresponds to the hypothetical first-order disappearance rate constant ($k = 0.693/t_{1/2}$) if Δt is much smaller than $t_{1/2}$.

HELLMAN *et al.* (1970) considered the mean concentration of cortisol during the interval and used the expression.

$$\text{Corrected cortisol level} = c_\tau - c_{(\tau-\Delta t)} + \frac{c_\tau + c_{(\tau-\Delta t)}}{2}\,\frac{\Delta t}{\overline{T}}$$

where $\overline{T} = 1/k = 1.44\, t_{1/2}\,.$

They calculated the amount of cortisol introduced into circulation in the interval Δt by multiplying the corrected level by V, the volume of distribution of the hormone. V and $t_{1/2}$ were estimated from the curve of disappearance of labeled cortisol from plasma, fit by a single exponential function.

As indicated in Chapters 1 and 2, all rates characterizing the exchange of material in a system of pools can be estimated if the production rates of each of the pools involved and the transfer factors between each pair of pools are known. Since we are assuming that the value of a ϱ factor is independent of the rates of production of the hormone, it follows that the basic problem in the estimation of average rates characterizing a system is to determine average production rates of each pool.

The average production rate of a circulating hormone ($\overline{PR}$) during the period of study can be estimated from measurements of the concentration of the hormone in blood, using the concept of a constant MCR. The MCR of a circulating hormone under non-steady-state conditions represents the ratio of the rate at which the hormone is irreversibly removed from circulation and the concentration of the hormone in blood (GURPIDE, 1972), i.e.,

$$MCR = \frac{v_{irr}}{c} = \frac{\varphi\, v_T}{c} = \frac{\varphi\,(v_{irr} + w_{out})}{c} \tag{5.9}$$

where φ is the fraction of the hormone leaving the pool irreversibly (Eq. 3.26), v_T is the total rate of exit, and v_{irr} is the rate of irreversible removal of the hormone from circulation. The rate w_{out}, which corresponds to the exit of material that is bound to return to circulation (or to the space of fast initial distribution of intravenously injected tracer of the hormone), may not be identical to the rate w_{in}, at which the hormone returns to circulation at any given instant, if the system is not at the steady state. It should be noted, however, that all material included in the rate w_{out} eventually will reenter the pool and will be included in the rate w_{in}.

An expression for $\overline{PR}$ (average production rate) is derived as follows:

$$V\frac{dc}{dt}=PR+w_{in}-(w_{out}+v_{irr})=PR+w_{in}-\frac{MCR}{\varphi}c\,.$$

Integration between t_2 and t_1,

$$V(c_{t_2}-c_{t_1})=\int_{t_1}^{t_2}PR\,dt+\int_{t_1}^{t_2}w_{in}\,dt-\frac{MCR}{\varphi}\int_{t_1}^{t_2}c\,dt \qquad (5.10)$$

where

$$\int_{t_1}^{t_2}w_{in}\,dt=\int_{t_1-\bar{t}_{11}}^{t_2-\bar{t}_{11}}w_{out}\,dt$$

when $\bar{t}_{11}$, the average time of transit of the compound outside the pool in one cycle (Eq. 3.32), is small in comparison to the interval t_2-t_1. From the definition of φ,

$$\int_{t_1-\bar{t}_{11}}^{t_2-\bar{t}_{11}}w_{out}\,dt=\int_{t_1-\bar{t}_{11}}^{t_2-\bar{t}_{11}}(1-\varphi)(w_{out}+v_{irr})\,dt=\frac{1-\varphi}{\varphi}MCR\int_{t_1-\bar{t}_{11}}^{t_2-\bar{t}_{11}}c\,dt\,.$$

Equation 5.10 is then equivalent to the following:

$$V(c_{t_2}-c_{t_1})=\overline{PR}(t_2-t_1)+\frac{1-\varphi}{\varphi}MCR\int_{t_1-\bar{t}_{11}}^{t_2-\bar{t}_{11}}c\,dt-\frac{MCR}{\varphi}\int_{t_1}^{t_2}c\,dt$$

or

$$\overline{PR}=V\frac{c_{t_2}-c_{t_1}}{t_2-t_1}+\frac{MCR}{\varphi(t_2-t_1)}\left[\int_{t_1}^{t_2}c\,dt-(1-\varphi)\int_{t_1-\bar{t}_{11}}^{t_2-\bar{t}_{11}}c\,dt\right]. \qquad (5.11)$$

MCR, V and φ can be estimated from tracer data. The average time lag $\bar{t}_{11}$ can also be measured from the same experimental data (Eq. 3.32). The integrals $\int_{t_1}^{t_2}c\,dt$ and $\int_{t_1-\bar{t}_{11}}^{t_2-\bar{t}_{11}}c\,dt$ can then be estimated from the appropriate areas under the concentration curve.

A simpler formula to calculate $\overline{PR}$ is applicable when the interval t_2-t_1 is sufficiently large. In this case, the two integrals in Eq. 511 become practically equal and, consequently,

$$\overline{PR}=V\frac{c_{t_2}-c_{t_1}}{t_2-t_1}+\frac{MCR}{t_2-t_1}\int_{t_1}^{t_2}c\,dt\,. \qquad (5.12)$$

Equation 5.12 also results from Eq. 5.11 when there is negligible reentry of the hormone into circulation ($\varphi \cong 1$).

Note 5.1

The peel-off and computer procedures for the estimation of multiexponential functions describing the disappearance of the tracer from circulation is illustrated with data obtained after intravenous injection of $^3H-E_2$ (LONGCOPE and WILLIAMS, 1974). This example will serve to evaluate the influence of various curve-fitting methods upon the calculated values of parameters such as volume of distribution and metabolic clearance rates.

Table 5.4. Concentrations of $^3H-E_2$ in plasma following an intravenous injection of $^3H-E_2$ to a nonpregnant young woman

Data point number	1	2	3	4	5	6	7	8	9	10	11	12
t: Time post-injection (min)	3	6	10	13	30	45	60	120	180	240	270	300
c*: Fraction of dose per liter of plasma (liter^{-1})×100	5.4	3.8	2.1	1.5	0.86	0.40	0.34	0.17	0.09	0.074	0.059	0.056

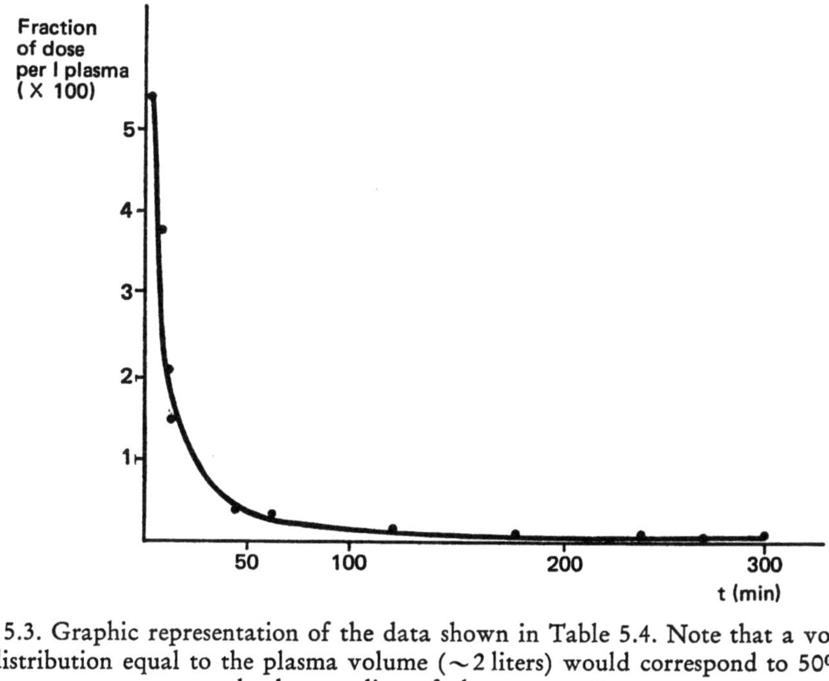

Fig. 5.3. Graphic representation of the data shown in Table 5.4. Note that a volume of distribution equal to the plasma volume (~2 liters) would correspond to 50% of the dose per liter of plasma at t=0

The experimentally determined concentrations of the tracer, corrected for losses occurring during the isolation procedures and expressed as percentage of the injected dose present in 1 liter of plasma, are shown in Table 5.4. A linear-linear plot of these concentrations versus time is shown in Fig. 5.3, and a semilogarithmic plot of the same data is shown in Fig. 5.4 (black circles).

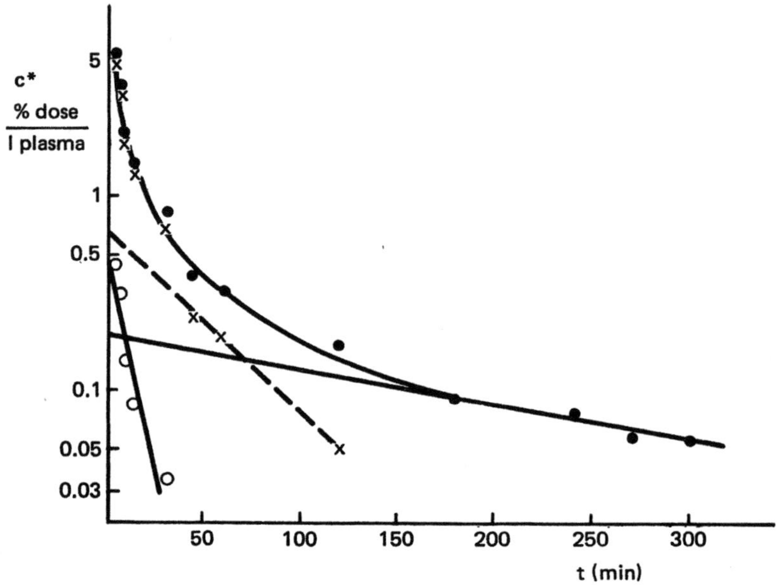

Fig. 5.4. Semilogarithmic plot of data shown in Table 5.4 (black circles). The three lines correspond to the exponential terms into which the tracer concentration function can be resolved by the "peel-off" procedure.

"Peel-off" Method

Visual inspection of the graph in Fig. 5.4 is used to decide how many of the late points in the graph can be validly fitted by a straight line. In making this decision it is useful to consider the errors associated with the datum points, both in regard to concentration of radioactivity and time of sampling. A straight line is fitted to the last four points and the values of D_n and α_n in the equation corresponding to this line ($\ln c^* = -\alpha_n t + \ln D_n$) are calculated. This can be achieved by using a ruler and fitting by eye the best line. Then, the value of D_n is determined as the intercept with the c^* axis and α_n is calculated from the value of the $t_{1/2}$ corresponding to this line ($\alpha_n = 0.693 \div t_{1/2}$). However, fitting of the straight line by the least-squares method (Selby, 1972) is preferable. Table 5.5 shows these calculations made with a programmable desk calculator. Note that the calculations are based on natural logarithms (although decimal logarithms could have been used). In order to

Table 5.5. Calculation of the last exponential term ($D_n\, e^{-\alpha_n t}$) of the function $c^* = \sum_{i=1}^{n} D_i\, e^{-\alpha_i t}$

t	c*	y = ln c*	$D_n\, e^{-\alpha_n t}$
180	9.0×10^{-4}	−7.0131	$y = -0.00418\, t - 6.2509$
240	7.4×10^{-4}	−7.2088	$\alpha_n = 0.00418;\ t_{1/2} = \frac{0.693}{\alpha_n} = 166$ minutes
270	5.9×10^{-4}	−7.4354	$\ln D_n = -6.2509$
300	5.6×10^{-4}	−7.4875	$D_n = 0.0019$

check for gross errors in the results, the calculated line is plotted in Fig. 5.4. Two points are sufficient to determine the line, e.g. ($c^* = D_n$, $t = 0$) and ($c^* = D_n/2$, $t = t_{1/2}$). The adequacy of the calculated line is apparent.

The steps necessary to calculate the parameters of the $(n-1)^{th}$ exponential ($D_{n-1}\, e^{-\alpha_{n-1} t}$) are shown in Table 5.6. The first two columns show experimental data, the third indicates the extrapolation values of the "tail" of the curve for early times, and the fourth gives the difference between the two preceding columns. These values are than plotted (Fig. 5.4, crosses) in order to decide, by visual inspection, the number of points to be used for the next straight line. The results of the application of the least-squares method to the points 45, 60, and 120 minutes are given in Table 5.6 and plotted in Fig. 5.4 (broken line).

The $(n-2)$ th exponential term is calculated following the same procedure. Table 5.7 lists the differences between the experimental data corresponding to the 3 – 30 minute samples and the values calculated from the last two

Table 5.6. Calculation of the second to last exponential term ($D_{n-1}\, e^{-\alpha_{n-1} t}$)

t	c*	$D_n\, e^{-\alpha_n t}$	$c^* - D_n\, e^{-\alpha_n t}$	$y = \ln (c^* - D_n\, e^{-\alpha_n t})$
3	54×10^{-3}	1.9×10^{-3}	52.1×10^{-3}	
6	38×10^{-3}	1.9×10^{-3}	36.1×10^{-3}	
10	21×10^{-3}	1.8×10^{-3}	19.2×10^{-3}	
13	15×10^{-3}	1.8×10^{-3}	13.2×10^{-3}	
30	8.6×10^{-3}	1.7×10^{-3}	6.9×10^{-3}	
45	4.0×10^{-3}	1.6×10^{-3}	2.4×10^{-3}	−6.0322
60	3.4×10^{-3}	1.5×10^{-3}	1.9×10^{-3}	−6.2659
120	1.7×10^{-3}	1.2×10^{-3}	0.5×10^{-3}	−7.6009

$$y = -0.0213 - 5.0357$$

$$\alpha_{n-1} = 0.0213;\ t_{1/2} = \frac{0.693}{\alpha_{n-1}} = 32.6 \text{ min}$$

$$\ln D_{n-1} = -5.0357$$

$$D_{n-1} = 0.0065.$$

Table 5.7. Calculation of the antepenultimate exponential term ($D_{n-2}\,e^{-\alpha_{n-2}t}$)

Time	c*	(a) $D_{n-1}\,e^{-\alpha_{n-1}t}$	(b) $D_n\,e^{-\alpha_n t}$	(d) $c^* - (a+b)$	$y = \ln d$
3	54×10^{-3}	6.1×10^{-3}	1.9×10^{-3}	46×10^{-3}	−3.0791
6	38×10^{-3}	5.7×10^{-3}	1.9×10^{-3}	30.4×10^{-3}	−3.4933
10	21×10^{-3}	5.3×10^{-3}	1.8×10^{-3}	13.9×10^{-3}	−4.2759
13	15×10^{-3}	4.9×10^{-3}	1.8×10^{-3}	8.3×10^{-3}	−4.7915
30	8.6×10^{-3}	3.4×10^{-3}	1.7×10^{-3}	3.5×10^{-3}	−5.6550

$$y = -0.0917\,t - 3.1214$$

$$\alpha_{n-2} = 0.0917;\; t_{1/2} = \frac{0.693}{\alpha_{n-2}} = 7.6\ \text{min}$$

$$\ln D_{n-2} = -3.1214$$

$$D_{n-2} = 0.0445.$$

exponential terms. These differences (fifth column in Table 5.7) are plotted in Fig. 5.4 (open circles). After visual inspection, a straight line is chosen to fit all these points. The values of the slope and intercept of this line which completes the peel-off procedure, are shown in Table 5.7 and illustrated in Fig. 5.4.

The first row in Table 5.8 shows the three exponential terms calculated in this manner. Obviously, some arbitrary decisions have been made during these calculations. For instance, a different choice of points during the application of the peel-off procedure could have led to a different number of exponential terms. Thus, in Fig. 5.4, rather than fitting one straight line to all the open circles, those at 13 and 30 minutes could have been used to determine the straight line corresponding to a $(n-3)$th term and consequently, another exponential term would be generated. The resulting function is listed in the second row of Table 5.8. Moreover, it is a matter of choice as to whether the last four or the last five data points in the original curve (black circles, Fig. 5.4) should be considered to be aligned. If five points are included in the tail of the curve, a different function results, as shown in the third row of Table 5.8. As seen in this table, not only the parameters of the calculated functions but also the values of volume of distribution, metabolic clearance rates and rate constants of exit of the tracer from circulation are strongly affected by the decisions made during the application of the peel-off method.

Computer Method (R. J. Bogumil)

A multiexponential approximation of the data in Table 5.4 has been performed using a digital computer-implemented procedure similar to that outlined in Chapter 3. The particular program employed is based on Mar-

Table 5.8. Calculation of parameters of the function $c^* = \Sigma_{i=1}^{n} D_i e^{-\alpha_i t}$ from the data in Table 5.4 using graphic and computer methods

Method	No. of exponential terms	Calculated parameters ($\times 10^3$) D_1 (l^{-1})	α_1 (min^{-1})	D_2 (l^{-1})	α_2 (min^{-1})	D_3 (l^{-1})	α_3 (min^{-1})	D_4 (l^{-1})	α_4 (min^{-1})	V (l)	MCR (l day^{-1})	$-\left(\frac{dc^*}{dt}\right) \times 10^3$ (l^{-1} min^{-1})
"Peel-off"												
Four points in tail	3	44.5	91.7	6.5	21.3	1.9	4.2	—	—	18.9	1160	4.2
	4 [a]	88.8	299	16.1	50.8	6.5	21.3	1.9	4.2	8.8	1050	27.5
Five points in tail	3	86.4	311	24.8	53.0	3.1	6.0	—	—	9.0	1140	28.1
Computer												
Weighted	2	48.1	88.9	4.17	7.3	—	—	—	—	19.2	1290	4.3
	3 [a]	78.1	239	17.5	42.9	2.6	5.4	—	—	10.2	1180	19.4
	4	75.0	282	22.6	67.3	5.3	12.9	0.4	0	9.7	0	22.8
Unweighted	2	76.4	170	9.2	13.9	—	—	—	—	11.7	1300	13.1
	3	75.3	177	10.5	20.4	0.7	0.5	—	—	11.6	630	13.6
	4	75.3	177	8.9	20.2	1.4	23.5	0.8	1.2	11.6	900	13.6

[a] Function considered to be best-fitting the data in each group of calculations.

quardt's algorithm (1963) and represents a minor modification of the programs available from the SHARE proram library agency, from Dr. R. DELL, Columbia University, New York, or other sources. The program computes values for the parameters (D's and α's) of the multiexponential function that minimize the sum of squared deviations with respect to the data. Each datum is assigned a weight, as described in Chapter 3. In particular, the weight may be uniform and equal to 1, i.e., the data points may be used "unweighted". The initial estimates were obtained by crude graphic analysis. Requirements on the precision of initial estimates are less stringent in the application of Marquardt's algorithm than is the case with the classic Newton-Gauss procedure described in Chapter 3.

The sum of squares (weighted or unweighted) of the difference between predicted and experimental values and the asymptotic variance estimates of the parameters were computed for functions of one to four exponential terms. This information is used to choose the function that "best" describes the data. The reader may again note that somewhat arbitrary and restrictive decisions must be employed to accomplish any such selection (between a three- or a four-exponential function in the present example). In each instance, the values of V, MCR, $(dc^*/dt)_0$ and others were calculated. Moreover, the actual area calculated from the data points by the trapezoidal rule (without extrapolation) was compared with the area calculated from the multiexponential function extrapolated to $t=0$ and $t=\infty$. The ratio between nonextrapolated and extrapolated areas serves to estimate the extent to which extrapolation influences the calculations of MCR. Clearly, the area under the measured points provides an upperbound on the MCR. As it happens, a lowerbound can not be provided because of uncertainty in the extrapolation.

Information on availability of this IBM 360, FORTRAN language program can be obtained upon request.

Table 5.8 shows the value of the parameters calculated using the data listed in Table 5.4, either weighted (assuming constant relative error) or unweighted (constant absolute error), for two, three, or four exponential terms.

Discussion of Results

Examination of the results included in Table 5.8 reveals the dependence of the calculated parameters upon the procedures followed for the analysis of the data. The following comments are pertinent:

1. The subjective decisions made during the graphic analysis of the data results in values of V which differ by a factor of two, and in even greater uncertainties in the values of the derivatives at $t=0$. In this example, MCR's are less affected. The variability of the results obtained by the graphic method depends entirely on the decision concerning the number of points to be ap-

proximated by each straight line as all regression lines were obtained with a desk calculator.

2. The results derived from the application of the computer method were strongly affected by weighting the data, the estimated MCR being significantly lower when uniform weights were assigned. Since in this example approximately constant relative error in the data is to be expected, uniform weighting is not appropriate. Note that in the "peel-off" method the data are in effect, weighted for constant relative error since the curve is divided in portions and approximated by least-squares straight line segments in semilog coordinates.

3. The three-exponential function obtained by computer analysis of the weighted data may be judged preferable to the function with four-exponential terms, as evidenced by lower variance estimates of the calculated parameters. The reader may note that increasing the number of exponential terms does not necessarily lead to a convergence to the true (unknown) values for V and MCR. This observation can be easily understood if it is considered that parameters such as V and MCR depend heavily on the extrapolation of the function to $t=0$ and $t=\infty$. In particular, a poor estimate for D_1 will significantly influence V and small absolute errors in α_n can affect the estimated value of the MCR. In fact, the four-exponential function fit to weighted data gave $\alpha_4=0$ and $MCR=0$.

The effect that omission of some data points may have on the calculated parameters is illustrated in Table 5.9. This table is particularly useful in showing the importance of early sampling in the estimation of volumes of distribution of the tracers.

Table 5.9. Effect of suppression of some experimental points from Table 5.4 [a]

Data points omitted (No.)	No. of exponential terms	V (liter)	MCR ($l\ day^{-1}$)	$-(dc^*/dt)_0$ ($l^{-1}\ min^{-1}$)
None	3	10.2	1180	19.4
1	2	25.8	1350	2.3
	3	4.3	980	83.1
1, 2	2	31.8	1420	1.6
12	3	9.9	1190	21.0
11, 12	2	13.7	1290	9.2
	3	9.9	1200	21.3
Odd numbers	2	11.3	1230	13.2
	3	10.1	1130	16.7
Even numbers	2	21.2	1260	3.0
	3	9.1	1140	29.1

[a] Multiexponential functions were calculated by the computer method using weighted data corresponding to the points listed in Table 5.4 less those indicated in the first column of this table. The number of exponential terms in each function was chosen by quality of fit. When the choice between two or three exponential terms was not obvious, values for V, MCR, and $(dc^*/dt)_0$ corresponding to both functions are shown.

Table 5.10. Sensitivity of parameter estimates to error in data points [a]

Procedure	Points deleted (No.)	"Error" added %	"Error" added Pattern	No. of exponential terms giving best fit	% Error in V	% Error in MCR	% Error in $\left(\frac{dc^*}{dt}\right)$
Peel-off	None	0	—	3	13	3.8	−26
Computer	None	0	—	4	0	0	0
		±2	↑↓	3	− 0.6	2.2	− 0.7
			↓↑	3	10	2.8	−22
		±5	↑↓	3	−10	1.4	24
			↓↑	3	19	3.8	−33
	1	0	—	4	0	0	0
		±2	↑↓	3	13	3.6	−26
			↓↑	3	20	5.2	−30
		±5	↑↓	3	9	2.5	−22
			↓↑	3	25	6.4	−33
	11, 12	0	—	4	0	0	0
		±2	↑↓	3	− 2.7	3.4	4.7
			↓↑	3	9	4.0	19
		±5	↑↓	3	−12	2.6	31
			↓↑	3	18	3.6	−32
	Even numbers	0	—	3	1.7	49	0
		±2	↑↓	3	−16	45	57
			↓↑	3	13	51	−22
		±5	↑↓	2	7.5	52	−22
			↓↑	3	26	54	−40

[a] *True function:* $c^* = 0.0888\,e^{-0.2986\,t} + 0.0161\,e^{-0.0508\,t} + 0.0065\,e^{-0.0213\,t} + 0.0019\,e^{-0.0042\,t}$. Values of c* at 3, 6, 10, 13, 30, 45, 60, 120, 180, 240, 270, and 300 minutes were calculated. Functions fitting these "simulated data points" were obtained by the peel-off and computer methods. Differences between the theoretical values of V, MCR, and $(dc^*/dt)_0$ and the corresponding values calculated from these functions are shown in the first two rows. Curve-fitting by computer was also applied to calculate multiexponential functions when some of the original simulated datum points were eliminated or were changed by alternately adding and subtracting a fixed error to each point. The arrows in the table indicate whether the error was added (↑↓) or subtracted (↓↑) to or from the first of the data points considered. Note that no more than three exponential terms can be fitted to six data points.

In order to evaluate the sensitivity of the calculated parameters to experimental errors in the data, it was assumed that the function of disappearance of the tracer obeyed the four-exponential function, listed in the second row of Table 5.8, which was obtained by the peel-off procedure. Values of simulated tracer concentrations at the times of sampling shown in Table 5.4 were generated using this function.

The calculated "data" points were first subjected to the "peel-off" procedure. It is interesting to note that only three-exponential terms were revealed

(Table 5.10). Furthermore, the agreement between true and calculated values for V, MCR, and $(dc^*/dt)_0$ were not completely satisfactory. The calculations by computer, however, yielded exactly the same four-exponential function from which the data were generated. As illustrated by this example, the computer optimization is clearly superior to the peel-off method in fitting points generated from a multiexponential function when the number of terms is moderate (or known *a priori*). It should be realized, however, that in biologic systems the true function of disappearance of the tracer may be a multiexponential function with a large number of terms or may contain nonexponential terms.

Table 5.10 also shows the results obtained by simulating various levels of measurement error. Whereas it is feasible to conduct extensive computer simulations in which the magnitude and sign of the error added to each point is assigned by a random process, a much simpler procedure, adequate for the present exposition, is merely to consider the effect of deviates of fixed magnitude (e.g., 2%, 5%) and alternating signs. In examples (such as the present case) in which a small number of observations (i.e., less than ten) are used to estimate a particular parameter (D's or α's) even randomly distributed errors are likely to exhibit some bias. The approach outlined above artificially minimizes these problems.

Note that addition of only 2% "error" to each point leads to a reduction of the number of identifiable exponentials from four to three. This result serves to illustrate once more the danger of drawing conclusions about the system on the basis of the number of exponential terms detected in the curve of disappearance of the tracer from circulation.

In this example, the parameters most sensitive to error in the data were the derivative of the function at $t = 0$ and the volume of distribution. The MCR was less affected by the same errors. It is remarkable, however, that a 50% overestimation of MCR results when only six data points are used, even when no errors are assigned to these points (Table 5.10, even numbers deleted). It should be noted that the error in MCR is due to differences in the extrapolated area since the fit of the three-exponential function is extremely good, as shown by the very small deviations (ranging between 0 and 10^{-4} relative error) at each data point.

It is then pertinent to demonstrate (see below) that an error in the estimate of MCR is introduced when the number of exponential terms in the approximate function (calculated by minimizing the sum of squared deviations) is lower than the number of terms in the true function. This result is of interest because the number of exponential terms in the circulating tracer decay function may be known to be very large from theoretical considerations, whereas the number of blood samples analyzed is limited.

By the arguments which follow it can be shown that if a series of values generated by the function $f(t) = \sum_{i=1}^{n} D_i e^{-\alpha_i t}$ are fitted by the least-squares method with a function $\hat{f}(t) = \Sigma_{j=1}^{m} \hat{D}_j e^{-\hat{\alpha}_j t}$ (α_j's with distinct real parts) and $m < n$, then $\Sigma_{i=1}^{n} (D_i/\alpha_i) \neq \Sigma_{j=1}^{m} (\hat{D}_j/\hat{\alpha}_j)$, i.e., $\widehat{MCR} \neq MCR$.

It follows from the first of Eqs. 3.46 a that

$$\sum_{k=1}^{L} \left[\sum_{j=1}^{m} \hat{D}_j e^{-\hat{\alpha}_j t_k} - \sum_{i=1}^{n} D_i e^{-\alpha_i t_k} \right] e^{-\hat{\alpha}_l t_k} = 0 \qquad l = 1, \ldots, m$$

or, equivalently, that

$$\sum_{k=1}^{L} \left[\sum_{j=1}^{m} \hat{D}_j e^{-(\hat{\alpha}_j + \hat{\alpha}_l) t_k} - \sum_{i=1}^{n} D_i e^{-(\alpha_i + \hat{\alpha}_l) t_k} \right] = 0 \qquad l = 1, \ldots, m.$$

If error-free continuous experimental data were obtained over the interval $t_1 = F$ to $t_L = L$, this equation becomes

$$\int_{F}^{L} \left[\sum_{j=1}^{m} \hat{D}_j e^{-(\hat{\alpha}_j + \hat{\alpha}_l) t} - \sum_{i=1}^{n} D_i e^{-(\alpha_i + \hat{\alpha}_l) t} \right] dt = 0 \qquad l = 1, \ldots, m$$

or
$$\sum_{j=1}^{m} \frac{\hat{D}_j}{\hat{\alpha}_j + \hat{\alpha}_l} \left(e^{-(\hat{\alpha}_j + \hat{\alpha}_l) F} - e^{-(\hat{\alpha}_j + \hat{\alpha}_l) L} \right) - \sum_{i=1}^{n} \frac{D_i}{\alpha_i + \hat{\alpha}_l} \left(e^{-(\alpha_i + \hat{\alpha}_l) F} - e^{-(\alpha_i + \hat{\alpha}_l) L} \right) = 0$$
$$l = 1, \ldots, m.$$

If $t_1 = F = 0$ and $t_L = L = \infty$, then

$$\sum_{i=1}^{n} \frac{D_i}{\alpha_i + \hat{\alpha}_l} - \sum_{j=1}^{m} \frac{\hat{D}_j}{\hat{\alpha}_j + \hat{\alpha}_l} = 0 \qquad l = 1, \ldots, m.$$

Therefore,

$$\sum_{i=1}^{n} \frac{D_i}{\alpha_i} \neq \sum_{j=1}^{m} \frac{\hat{D}_j}{\hat{\alpha}_j}$$

unless one of the $\hat{\alpha}_l = 0$.

Under less ideal circumstances (e.g., when the duration of the experiment is limited, the data are not recorded continuously and measurement errors are present) the least-squares criterion will yield the theoretical (true) $\Sigma_{i=1}^{n} (D_i/\alpha_i)$ or MCR only by fortunate coincidence. The magnitude of the discrepancy between MCR and $\widehat{MCR}$ can be small or large, as seen in the example shown in Table 5.10. Thus, when a three exponential term function was fitted by the least-squares method to the 12 simulated data points generated by the four-exponential-term functions, the relative error in the calculated MCR was only 2.7%.

Note 5.2

The values of w_{E_1} and w_{E_2} in Fig. 1.5 were calculated using functions describing the concentration of intravenously injected $^3H-E_1$ and $^3H-E_2$ [c^* (fraction of injected dose remaining per liter of plasma) $= D_1 e^{-\alpha_1 t} + D_2 e^{-\alpha_2 t} + D_3 e^{-\alpha_3 t}$, t in minutes] (Longcope and Williams, 1974):

Tracer	D_1	α_1	D_2	α_2	D_3	α_3
$^3H-E_1$ injected	0.058	0.3736	0.0067	0.0304	0.0012	0.0029
$^3H-E_2$ injected	0.071	0.1996	0.0140	0.0292	0.0019	0.0042

From these data, and using average concentration values for E_1 and E_2 in plasma (NAGAI and LONGCOPE, 1971), the following values were estimated:

	E_1	E_2
$V = 1/\Sigma D_i$ (liter plasma)	15	11.5
$-\left(\frac{dc}{dt}\right)_0^* = \Sigma D_i \alpha_i \left(\frac{1}{\text{liter} \times \text{min}}\right)$	0.0219	0.0146
$k_{total} = -\left(\frac{dt}{dt}\right)_0^* V \left(\frac{1}{\text{min}}\right)$	0.3282	0.1676
c (ng/liter plasma)	30.1	25.6
$M = c\,V$ (ng)	451	294
$r_{total} = k_{total} M \left(\frac{\text{nmole}}{\text{day}}\right)$	790	261
$w_{E_1} = r_{total, E_1} - (v_{E_1E_2} + v_{E_1O}) \left(\frac{\text{nmole}}{\text{day}}\right)$	524	—
$w_{E_2} = r_{total, E_2} - (v_{E_2E_1} + v_{E_2O}) \left(\frac{\text{nmole}}{\text{day}}\right)$	—	89

Note 5.3

The value of w_{E_1S} in Fig. 1.7 was calculated from data reported by LONGCOPE (1972), i.e.,

$$c_{E_1S}^{^3H} = D_1\, e^{-\alpha_1 t} + D_2\, e^{-\alpha_2 t} = 0.121\, e^{-0.2249t} + 0.025\, e^{-0.0035t}$$

where $c_{E_1S}^{^3H}$ is the concentration of the injected tracer (fraction of dose per liter of plasma) and t is expressed in minutes. Plasma levels of E_1S in men have been reported by LORIAUX *et al.* (1971). Thus,

$$V = 1/\Sigma D_i = 6.8 \text{ liter}$$

$$-\left(\frac{dc}{dt}\right)_0^* = \Sigma D_i \cdot \alpha_i = 0.0273 \quad \frac{1}{\text{min} \times \text{liter}}$$

$$k_{total} = -\left(\frac{dc}{dt}\right)_0^* V = 0.1869 \text{ min}^{-1}$$

$$c_{E_1S} = 1.22 \text{ nmole}$$

$$M = c\,V = 8.3 \text{ nmole}$$

$$r_{total} = k_{total}\, M = 2237 \text{ nmole/day}$$

$$w_{E_1S} = r_{total} - (v_{E_1S,E_1} + v_{E_1S,E_2} + V_{E_1S,0}) = 1976 \text{ nmole/day}.$$

Similarly,

$$w_{E_1} = k_{total\ E_1} M_{E_1} - (v_{E_1E_1S} + v_{E_1E_2} + v_{E_10})$$

and

$$w_{E_2} = k_{total\ E_2} M_{E_2} - (v_{E_2E_1S} + v_{E_2E_1} + v_{E_20}).$$

Chapter 6

Rates of Secretion and Metabolism of Hormones Estimated from Specific Activities of Urinary Metabolites

The original methods to measure secretory rates were based on the concept that the daily production of a hormone could be estimated from the amount of one of its metabolites excreted in urine during 24 hours if the relation that exists between the rate of release of the hormone into circulation and the rate of excretion of the metabolite were known. This relation can be determined by administering a known dose of the hormone and measuring the total amount of the metabolite in urine in excess over the basal level of excretion.

BROWN (1957) estimated by this means the rate of secretion of estradiol in women. He injected intramuscularly 1 mg of the hormone and recovered from urine 16% of the administered dose as estradiol, estrone, and estriol. He then multiplied the values of "total estrogens" measured in other urine samples by 100/16 and found that approximately 300 μg/24 hr of estradiol were secreted at midcycle by ovulating women.

When isotopically labeled tracers became available, it was possible to repeat these experiments by injecting the radioactive hormone and measuring the fraction of the administered isotope that appears in the urinary metabolite. With this procedure, usually negligible amounts of exogenous hormone need be administered and the normal physiologic hormonal state is not altered. Although the same results were obtained by administering either a trace amount of radioactive estradiol or a load of unlabeled hormone, this is not necessarily the case for all compounds. For example, the fraction of endogenously produced dehydroisoandrosterone excreted as urinary dehydroisoandrosterone sulfate depends on the amount of the compound entering the systemic circulation (MACDONALD *et al.*, 1962). In cases like this, the use of tracers is necessary to avoid changes of the endogenous concentration of the compound in the subject under study.

A major advantage of the tracer method is that both the conversion of the secreted hormone to the urinary metabolite and the excretion of the metabolite can be measured in a single experiment.

The tracer method was introduced in the steroid field by PEARLMAN (1957), ULICK *et al.* (1958), and COPE and BLACK (1958) in their studies of progesterone, aldosterone, and cortisol secretion, respectively. The general applicability of the method was challenged, however, when it became apparent that urinary metabolites of steroids may derive from various secreted precursors rather than from a single hormone (VANDE WIELE *et al.*, 1963). In fact, multiplicity of precursors of urinary steroids is of common occurrence; very few steroid hormones have metabolites that are not derived from other secreted compounds. Some of these complex situations, however, can be resolved by using more than one isotopically labeled tracer, as is shown in this chapter.

Specific activity data provided by urinary metabolites during tracer experiments yield a wealth of information about rates at which precursors are produced and metabolic reactions occur. However, the interpretation of the tracer data requires exacting evaluation of the origin of each compound since the metabolites are end products of complex metabolic reactions and may be formed in more than one tissue.

This chapter describes procedures to estimate rates of hormone secretion and rates of metabolic reactions, including special situations associated with pregnancy.

A. Estimation of Hormone Secretion Rates

Case 1. Metabolites Derived from One Secreted Precursor (Fig. 6.1, e. g., Aldosterone)

The specific activity ($\bar{a}$) in dpm/mmole of a metabolite isolated from urine equals, by definition, the ratio of the amounts of isotope to the amount of endogenous metabolite excreted in urine during the time of collection. The amount of radioactivity in the metabolite equals the dose of tracer injected (in dpm) multiplied by the fraction (f^*) of the dose appearing in urine as the metabolite in question. On the other hand, the amount of endogenous metabolite excreted during the time (t) of urine collection, equals the rate of secretion of the hormone (Q) multiplied by t and by the fraction (f) of the endogenous hormone converted to the urinary metabolite. The following expression describes the specific activity of the metabolite:

$$\bar{a}\,(\text{dpm/mmole}) = \frac{\text{Dose (dpm)} \times f^*}{Q\,(\text{mmole/day} \times f \times t\,(\text{days})}\,. \tag{6.1}$$

If the fractions f^* and f in the above equation are identical, we have the simple expression

$$Q = \frac{\text{Dose}}{\bar{a} \times t} \tag{6.2}$$

stating that the secretion rate of the hormone equals the dose injected divided by the specific activity of the metabolite and by the time of urine collection.

The product of the specific activity of a metabolite and the time during which urine is collected becomes a constant when the amount of radioactivity

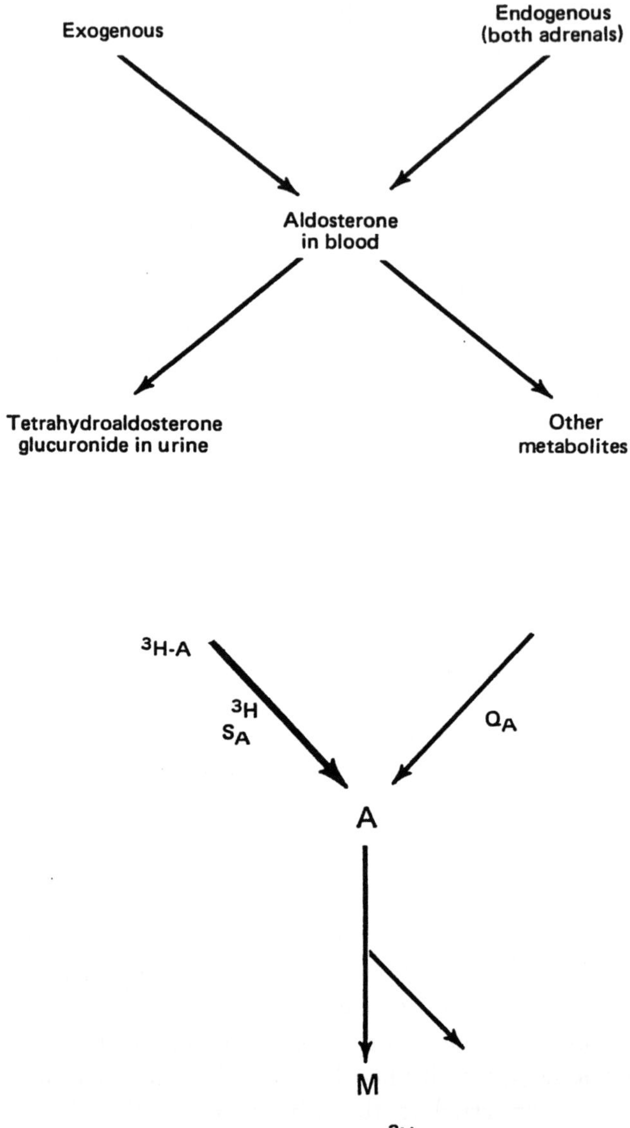

$$Q_A = \frac{S_A^{^3H}}{\bar{a}_M^{^3H} \times t}$$

Fig. 6.1. *Urinary metabolites derived from a single secreted precursor* (Case 1, e. g., aldosterone). The endogenously produced hormone (aldosterone) and the intravenously injected tracer are metabolized to many products among which urinary tetrahydroaldosterone glucuronide is selected. The simplest representation of this situation, in which the metabolite M derives solely from the secreted hormone A is shown together with the formula by which the rate of secretion of the hormone (Q_A, in nmole/day) can be estimated from the dose of tracer injected ($S_A^{^3H}$, in dpm), the adequately long time of urine collection (t) and the cumulative specific activity of the metabolite M ($\bar{a}_M^{^3H}$, in dpm/nmole)

(E) excreted as the labeled metabolite becomes constant. This statement is evident, since $\bar{a} = E/(\varepsilon \times t)$ or $\bar{a} \times t = E/\varepsilon$ where ε is the rate of excretion of the metabolite. Therefore,

$$\bar{a} \times t = \frac{E_{final}}{\varepsilon} = \frac{f^* \, Dose}{\varepsilon} = \text{constant} \tag{6.3}$$

when no more labeled metabolite is excreted. Consequently, urine collection can be terminated at any time after E becomes practically constant.

Assumptions

An assumption basic to the derivation of Eq. 6.1 is that the fates of tracer and tracee are identical ($f^* = f$). Since both secreted hormone and intravenously injected tracer are rapidly transported to the same anatomic space, viz. the right heart, this assumption seems reasonable. In general, it appears that the presence in the steroid molecule of heavier hydrogen ([^{3}H]) or carbon ([^{14}C]) atoms or the labeling of a protein hormone with isotopic iodine atoms ([^{125}I] and [^{131}I]) does not significantly affect the metabolism of the tracers. Isotope effects have been demonstrated, however, in some metabolic reactions of steroids (KARAVOLAS *et al.*, 1969).

It was also assumed in deriving Eq. 6.1 that the time t of urine collection was adequate to allow the complete excretion of labeled metabolite. Fulfillment of this condition is usually claimed when the amount of radioactivity excreted becomes negligible with respect to the amount already collected. Prolonged excretion of "negligible" amounts of radioactivity, however, may add up to a significant quantity of labeled metabolite. This potential source of error in the method is particularly apparent in the case of hormones such as progesterone, for which low recovery of the injected radioactivity is reported (ROMANOFF *et al.*, 1963).

The fraction converted to the urinary metabolite was assumed to remain constant during the entire experiment. If this were not the case, the metabolic fate of the injected labeled hormone might be different from that of the endogenous hormone secreted at a later time. To verify the constancy of this fraction, subjects can be injected with [^{3}H] and [^{14}C] (or [^{125}I] and [^{131}I])-labeled tracers of the same hormone at different times of the same day. If the fractional rates of the metabolic processes are constant, the isotope ratio of metabolites isolated from urine collected over a period of time adequate for complete excretion of radioactivity should equal the isotope ratio of the injected tracers (VANDE WIELE *et al.*, 1963; KELLY, 1970). The fraction of secreted hormone converted to the urinary metabolite may remain constant even if the secretion rate varies. If such is the case, the urinary isotope dilution method yields an average secretion rate over the period of urine collection.

The pathways by which the hormone is converted to the metabolite, and the extent of such conversions, are irrelevant to the estimation of secretory rates. Furthermore, since specific activities are used for the calculations, losses incurred during the extraction and isolation procedures do not affect the results.

Application

The single isotope method is used to measure the secretion rate of aldosterone, a hormone that appears to be the only secreted precursor of various metabolites, tetrahydroaldosterone among them (LARAGH and KELLY, 1964).

When the metabolite is also derived from another secreted precursor neither Eq. 6.1 nor Eq. 6.2 can be used. Note that in this case, the numerator in Eq. 6.1 is correct but the denominator fails to describe the contribution of the other precursors to the weight of the metabolite.

Case 2. Metabolites Derived from Two Secreted Precursors (Fig. 6.2, e. g., Dehydroisoandrosterone)

When metabolites derive from two secreted precursors, the calculation of secretion rates requires the administration of two tracers, one for each of the secreted precursors of the common metabolites, followed by measurement of the specific activities of two urinary metabolites.

The derivation of the expressions used to calculate secretion rates from isotopic data is based on the scheme shown in Fig. 6.2.

A mixture of tritiated tracer of compound A and [^{14}C]-labeled tracer of compound B is injected intravenously, and urine, collected for several days, is pooled and processed for the isolation of metabolites I and II. The specific activities of these metabolites, with respect to both [^{3}H] and [^{14}C], are then determined. If A and B are the only secreted precursors of I and II, it follows from the definition of a specific activity that

$$\bar{a}_{I}^{^3H} = \frac{\text{amt. of } ^3\text{H-I in pooled urine}}{\text{amt. of I in pooled urine}} = \frac{S_A^{^3H} \times f_{A\to I}^{^3H}}{(Q_A \times f_{A\to I} + Q_B \times f_{B\to I})\, t} \tag{6.4}$$

and

$$\bar{a}_{II}^{^{14}C} = \frac{\text{amt. of } ^{14}\text{C-II in pooled urine}}{\text{amt. of II in pooled urine}} = \frac{S_B^{^{14}C} \times f_{B\to II}^{^{14}C}}{(Q_A \times f_{A\to II} + Q_B \times f_{B\to II})\, t}\,. \tag{6.5}$$

As indicated in Fig. 6.2, $S_A^{^3H}$ and $S_B^{^{14}C}$ are the amounts of radioactivity injected with the ^{3}H-A and ^{14}C-B tracers, respectively. It is assumed that the weight of these tracers is negligible; if this is not the case, appropriate corrections should be made. The f factors represent fractions of tracers or endogenous compounds excreted as a particular metabolite. For instance, $f_{A\to I}^{^3H}$ indicates the fraction of tracer ^{3}H-A excreted as ^{3}H-I. Since the endogenously secreted compound A and the intravenously injected ^{3}H-A have a common starting point for their metabolic journey, viz. the right auricle, $f_{A\to I}$, the

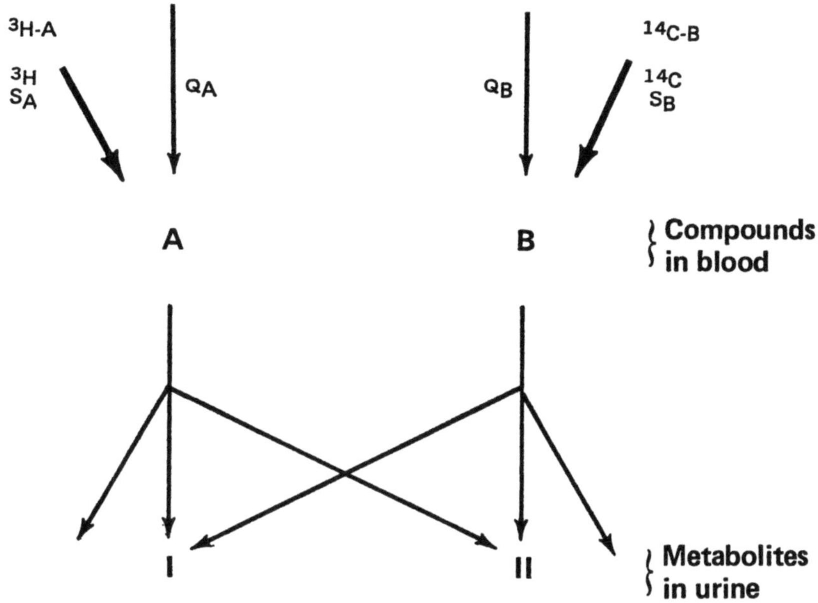

$$Q_A = \frac{S_A^{3H}}{t} \cdot \frac{\bar{a}_{II}^{14C} - \bar{a}_{I}^{14C}}{\bar{a}_{I}^{3H}\,\bar{a}_{II}^{14C} - \bar{a}_{II}^{3H}\,\bar{a}_{I}^{14C}} \qquad Q_B = \frac{S_B^{14C}}{t} \cdot \frac{\bar{a}_{I}^{3H} - \bar{a}_{II}^{3H}}{\bar{a}_{I}^{3H}\,\bar{a}_{II}^{14C} - \bar{a}_{II}^{3H}\,\bar{a}_{I}^{14C}}$$

Fig. 6.2. *Urinary metabolites derived from two secreted precursors* (Case 2, e. g., dehydroisoandrosterone and its sulfate). Urinary metabolites I and II are solely derived from compounds A and B, secreted at rates Q_A and Q_B. The pathways by which A and B are converted to I and II need not be postulated. If a mixture of tracers of A and B, labeled with different isotopes, in injected intravenously, the rates Q_A and Q_B can be estimated from the formulas given at the bottom of the Figure, provided that the denominator is significantly different from zero. Metabolic characteristics that lead to inadequate values of the denominator are discussed in the text

fraction of endogenously secreted A excreted as I, may be assumed to equal $f_{A\to I}^{3H}$. Hence, dividing numerator and denominator of Eq. 6.4 by $f_{A\to I}^{3H}$ we have

$$\bar{a}_I^{3H} = \frac{S_A^{3H}}{\left(Q_A + Q_B \dfrac{f_{B\to I}}{f_{A\to I}}\right) t} . \tag{6.6}$$

From similar arguments, it can be stated that $f_{B\to I}^{14C} = f_{B\to II}$ and therefore, Eq. 6.5 can be written in the following form:

$$\bar{a}_{II}^{14C} = \frac{S_B^{14C}}{\left(Q_A \dfrac{f_{A\to II}}{f_{B\to II}} + Q_B\right) t} . \tag{6.7}$$

The values of the f factors are obtained from isotopic data. For instance, from the definition of $f^{^3H}_{A\to I}$, it is apparent that this fraction equals the total amount of tritium excreted as metabolite I in urine ($E_I^{^3H}$) divided by the dose of ^{3}H-A injected. Therefore, $f_{A\to I} = f^{^3H}_{A\to I} = E_I^{^3H}/S_A^{^3H}$. Similarly, $f_{A\to II} = E_{II}^{^3H}/S_A^{^3H}$, $f_{B\to I} = E_I^{^{14}C}/S_B^{^{14}C}$, and $f_{B\to II} = E_{II}^{^{14}C}/S_B^{^{14}C}$. Consequently,

$$\frac{f_{B\to I}}{f_{A\to I}} = \frac{E_I^{^{14}C} \times S_A^{^3H}}{E_I^{^3H} \times S_B^{^{14}C}} = \frac{(^3H/^{14}C)\,\mathrm{inj}}{(^3H/^{14}C)_I} \tag{6.8}$$

and

$$\frac{f_{A\to II}}{f_{B\to II}} = \frac{E_{II}^{^3H}\; S_B^{^{14}C}}{E_{II}^{^{14}C}\; S_A^{^3H}} = \frac{(^3H/^{14}C)_{II}}{(^3H/^{14}C)_{\mathrm{inj}}}\,. \tag{6.9}$$

These are the ratios that appear in Eqs. 6.6 and 6.7. They are evaluated in terms of isotope ratios in the injected mixture and in the metabolite; data on total amounts of isotope excreted in urine are not required. Using these expressions, Eqs. 6.6 and 6.7 take the form

$$Q_A + \frac{(^3H/^{14}C)\,\mathrm{inj}}{(^3H/^{14}C)_I}\, Q_B = \frac{S_A^{^3H}}{\bar{a}_I^{^3H} t}\,, \tag{6.10}$$

$$\frac{(^3H/^{14}C)_{II}}{(^3H/^{14}C)\,\mathrm{inj}}\, Q_A + Q_B = \frac{S_A^{^3H}}{\bar{a}_{II}^{^{14}C} t}\,. \tag{6.11}$$

This system of two linear simultaneous equations in two unknowns, Q_A and Q_B, can be solved by any of the various methods described in elementary textbooks of algebra.

If the method of determinants is used, the intermediate steps are as follows:

$$Q_A = \begin{vmatrix} \dfrac{S_A^{^3H}}{\bar{a}_I^{^3H}} & \dfrac{(^3H/^{14}C)\,\mathrm{inj}}{(^3H/^{14}C)_I} \\ \dfrac{S_B^{^{14}C}}{\bar{a}_B^{^{14}C} t} & 1 \end{vmatrix} \cdot/\cdot \begin{vmatrix} 1 & \dfrac{(^3H/^{14}C)\,\mathrm{inj}}{(^3H/^{14}C)_I} \\ \dfrac{(^3H/^{14}C)_{II}}{(^3H/^{14}C)\,\mathrm{inj}} & 1 \end{vmatrix}$$

$$= \frac{S_A^{^3H}}{\bar{a}_I^{^3H} t} - \frac{S_B^{^{14}C}\,(^3H/^{14}C)_{\mathrm{inj}}}{\bar{a}_{II}^{^{14}C}(^3H/^{14}C)_I t} \cdot/\cdot \left(1 - \frac{(^3H/^{14}C)_{II}}{(^3H/^{14}C)_I}\right)$$

$$= \frac{S_A^{^3H}}{t}\, \frac{\bar{a}_{II}^{^{14}C}\,(^3H/^{14}C)_I - \bar{a}_I^{^3H}}{\bar{a}_I^{^3H}\, \bar{a}_{II}^{^{14}C}\,(^3H/^{14}C)_I} \cdot/\cdot \frac{(^3H/^{14}C)_I - (^3H/^{14}C)_{II}}{(^3H/^{14}C)_I}$$

$$= \frac{S_A^{^3H}}{t}\,(\bar{a}_{II}^{^{14}C} - \bar{a}_I^{^{14}C}) \cdot/\cdot\; \bar{a}_I^{^{14}C}\, \bar{a}_{II}^{^{14}C}\, [(^3H/^{14}C)_I - (^3H/^{14}C)_{II}]\,.$$

Finally,

$$Q_A = \frac{S_A^{^3H}}{t}\, \frac{\bar{a}_{II}^{^{14}C} - \bar{a}_I^{^{14}C}}{\bar{a}_I^{^3H}\, \bar{a}_{II}^{^{14}C} - \bar{a}_I^{^{14}C}\, \bar{a}_{II}^{^3H}}\,. \tag{6.12}$$

Similarly,

$$Q_B = \frac{S_B^{^{14}C}}{t}\, \frac{\bar{a}_I^{^3H} - \bar{a}_{II}^{^3H}}{\bar{a}_I^{^3H}\, \bar{a}_{II}^{^{14}C} - \bar{a}_I^{^{14}C}\, \bar{a}_{II}^{^3H}}\,. \tag{6.13}$$

Equations 6.12 and 6.13 provide solutions for the case involving two secreted compounds with common metabolites. Their application depends upon the availability of two metabolites that are entirely derived from the two secreted compounds in question. No assumptions are made about the paths by which the secreted compounds are converted to the urinary products, or about the sites of such conversions.

Of course, compounds A and B may be metabolized to products other than I and II. Thus, another pair of metabolites could have been used for the calculation of Q_A and Q_B.

The specific activities of I and II should be such that the denominators in Eqs. 6.12 and 6.13 do not equal 0 as it would occur if the specific activities of both metabolites were identical with respect to [^{3}H] as well as to [^{14}C], i.e., if $\bar{a}_I^{^3H} = \bar{a}_{II}^{^3H}$ and $\bar{a}_I^{^{14}C} = \bar{a}_{II}^{^{14}C}$. Under such conditions, calculation would be impossible.

Two metabolites may have identical specific activities if they are derived from a common metabolic precursor, e.g., $A \rightarrow C \rightarrow I + II$ and $B \rightarrow B \rightarrow C \rightarrow I + II$. The specific activities of metabolites I and II, derived from a single chemical intermediate C, may be different, however, if they are formed in different tissues, e.g., in the maternal and fetal organisms during pregnancy.

Identical specific activities of two metabolites may also result from rapid interconversion between them or between their precursors.

In addition, two metabolites may show identical specific activities if their secreted precursors contribute equally to each of them. Consider Eqs. 6.2 and 6.4 and the following two similar expressions:

$$\bar{a}_{II}^{^3H} = \frac{S_A^{^3H} \times f_{A \rightarrow II}}{(Q_A\, f_{A \rightarrow II} + Q_B\, f_{B \rightarrow II})\, t} \tag{6.14}$$

and

$$\bar{a}_I^{^{14}C} = \frac{S_B^{^{14}C}\, f_{B \rightarrow I}}{(Q_A\, f_{A \rightarrow I} + Q_B\, f_{B \rightarrow I})\, t} \,. \tag{6.15}$$

It is evident from these four equations that $\bar{a}_I^{^3H} = \bar{a}_{II}^{^3H}$ and $\bar{a}_I^{^{14}C} = \bar{a}_{II}^{^{14}C}$ if $f_{A \rightarrow I} = f_{A \rightarrow II}$ and $f_{B \rightarrow I} = f_{B \rightarrow II}$. When this situation is suspected, it is advisable to examine several metabolic products, since a third metabolite may show a different specific activity even if it is derived solely from A and B.

Equations 6.12 and 6.13 are also applicable to the situation illustrated by Fig. 6.3 in which one of the metabolites derives from only one precursor while the other receives contributions from two.

Although $\bar{a}_I^{^3H} = \bar{a}_{II}^{^3H}$ when $Q_B = 0$ (Fig. 6.2; Eqs. 6.4 and 6.14), $\bar{a}_I^{^{14}C}$ may differ from $\bar{a}_{II}^{^{14}C}$ as illustrated by Eqs. 6.5 and 6.15. In this case, the correct answer, e.g., $Q_B = 0$, may be obtained. The inherent error in the measurement of specific activities, compounded by the complexity of the expressions from which secretion rates are calculated, makes urinary methods inadequate for distinguishing between a nil secretion and a secretion rate that is small compared to the secretion rate of the other precursor. This limitation is particularly evident when the specific activities of the urinary metabolites tend to be equalized by an extensive interconversion between the precursors (KOWARSKI *et al.*, 1969).

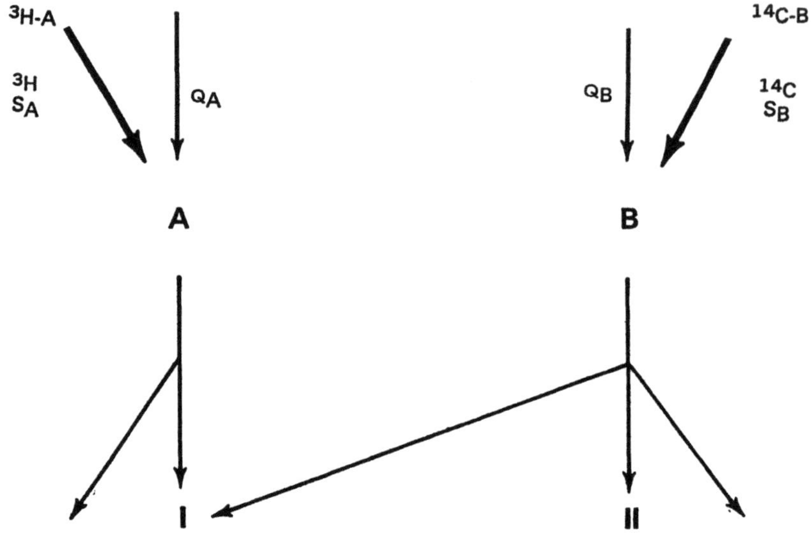

$$Q_A = \frac{S_A^{3H}}{t} \cdot \frac{\bar{a}_{II}^{14C} - \bar{a}_{I}^{14C}}{\bar{a}_{I}^{3H}\,\bar{a}_{II}^{14C} - \bar{a}_{II}^{3H}\,\bar{a}_{I}^{14C}} \qquad Q_B = \frac{S_B^{14C}}{\bar{a}_B^{14C}\; t}$$

Fig. 6.3. Special case ($\bar{a}_{II}^{3H} = 0$) of the situation shown in Fig. 6.2

Example

A normal woman received intravenously 40.5×10^6 dpm of [^{3}H]dehydroisoandrosterone (D) and 1.72×10^6 dpm of [^{14}C]dehydroisoandrosterone sulfate (DS). Urine was collected for 3 days following the injection and dehydroisoandrosterone glucuronide (DG) and DS were isolated from the pooled urine. The specific activities in these metabolites were the following: DG, 530,000 dpm [^{3}H]/μmole, 36,000 dpm [^{14}C]/μmole; DS, 720,000 dpm [^{3}H]/μmole, 14,700 dpm [^{14}C]/μmole. From these data, it could be estimated that the subject secreted 6.1 μmoles of D and 15.7 μmoles of DS daily (SANDBERG *et al.*, 1964).

Case 3. Metabolites Derived from Several Secreted Precursors (e. g., Testosterone)

The metabolic products of some hormones have more than two secreted precursors. The metabolites of testosterone, for instance, are known to derive from secreted androstenedione, dehydroisoandrosterone, and dehydroisoandrosterone sulfate as well as from testosterone (VANDE WIELE *et al.*, 1963). Urinary estrogens originate not only from secreted estrone and estradiol but also from secreted androgens, testosterone, and androstenedione (MACDONALD *et al.*, 1967). Another urinary metabolite, pregnanediol, is formed from progesterone as well as from pregnenolone and 20α-dihydroprogesterone, all of these precursors being secretory products of the adrenals or the gonads (ARCOS *et al.*, 1964).

To measure secretion rates of hormones with metabolites derived from multiple precursors, the administration of three or more tracers is necessary. It is not difficult to derive formulas for the calculation of secretion rates from the specific activities of urinary metabolites, regardless of the number of precursors involved. The procedures to be followed are essentially identical to those used to obtain Eqs. 6.4 to 6.13. The result is the general formula for *m* metabolites and hormones (GURPIDE and MANN, 1965):

$$\bar{a}_q^1 = E_q^1 \left(\frac{E_q^1}{S_1^*} Q_1 + \frac{E_q^2}{S_2^*} Q_2 + \cdots + \frac{E_q^m}{S_m^*} Q_m \right)$$

where S*, E, ā, Q, and t denote, respectively, dose injected, total isotope in the metabolite, specific activity of the metabolite, secretion rate of the hormone, and time of urine collection. The superscript attached to the ā's indicates the isotope by referring to the tracer administered. From this formula, it follows that

$$\frac{S_1^*}{\bar{a}_q^1 t} = Q_1 + \frac{S_1^*}{S_2^*} \frac{E_q^2}{E_q^1} Q_2 + \cdots + \frac{S_1^*}{S_m^*} \frac{E_q^m}{E_q^1} Q_m$$

$$\frac{S_1^*}{\bar{a}_q^1 t} = Q_1 + \frac{S_1^*}{S_2^*} \frac{\bar{a}_q^2}{\bar{a}_q^1} Q_2 + \cdots + \frac{S_1^*}{S_m^*} \frac{\bar{a}_q^m}{\bar{a}_q^1} Q_m$$

or, if both sides are multiplied by $\bar{a}_q^1 t / S_1^*$

$$\sum_{i=1}^{m} \frac{\bar{a}_q^i t}{S_i^*} Q_i = 1 \qquad q = 1, 2, \ldots, m .$$

For instance, if four compounds (A, B, C, D) are possible precursors of the metabolites of A, then tracers of these four compounds should be administered and specific activities of four metabolites (I, II, III, IV) should be measured. If only two isotopes are used to label the tracers (e.g., [^{3}H] and [^{14}C]), two consecutive experiments must be performed. Labeled A and B may be administered first while labeled C and D are injected later. The following expression, in determinant form, is obtained from the general formula given above and can be used to calculate the secretion rate of compound A:

$$Q_A = \frac{S_A^*}{t} \begin{vmatrix} 1 & \bar{a}_I^B & \bar{a}_I^C & \bar{a}_I^D \\ 1 & \bar{a}_{II}^B & \bar{a}_{II}^C & \bar{a}_{II}^D \\ 1 & \bar{a}_{III}^B & \bar{a}_{III}^C & \bar{a}_{III}^D \\ 1 & \bar{a}_{IV}^B & \bar{a}_{IV}^C & \bar{a}_{IV}^D \end{vmatrix} \cdot/\cdot \begin{vmatrix} \bar{a}_I^A & \bar{a}_I^B & \bar{a}_I^C & \bar{a}_I^D \\ \bar{a}_{II}^A & \bar{a}_{II}^B & \bar{a}_{II}^C & \bar{a}_{II}^D \\ \bar{a}_{III}^A & \bar{a}_{III}^B & \bar{a}_{III}^C & \bar{a}_{III}^D \\ \bar{a}_{IV}^A & \bar{a}_{IV}^B & \bar{a}_{IV}^C & \bar{a}_{IV}^D \end{vmatrix} .$$

Specifically, compounds A, B, C, and D may represent testosterone, androstenedione, dehydroisoandrosterone, and dehydroisoandrosterone sulfate. The four metabolites, which must differ in the value of at least one of their specific activities, may be testosterone glucuronide, androsterone glucuronide, dehydroisoandrosterone glucuronide, and dehydroisoandrosterone sulfate.

If an experiment of this kind is conducted, values for the rates of secretion of the four compounds can be estimated. However, the accumulation of errors in the measurement of specific activities and the variability of secretion rates and rates of metabolism during the long period of time necessary to conduct two successive experiments, limit the applicability of this experimental design.

B. Interpretation of Specific Activity Data Obtained from Labeled Urinary Metabolites

If a dose S_A^* dpm of a tracer of A is injected intravenously, the specific activity of a metabolite M, isolated from urine collected for a period t following the injection, is given by the expression $\bar{a}_M^A = f_{A\to M} S_A^*/(\varepsilon_M t)$, if ε_M, the daily amount of M excreted in urine, is constant and t is large enough so that $\bar{a}_M^A \times t$ is constant (Eq. 6.3). Since it can be stated in general that M is derived from precursors A, B, ..., J, which are secreted at rates $Q_A, Q_B, \ldots, Q_J$, respectively, it follows that $\varepsilon_M = f_{A\to M} Q_A + f_{B\to M} Q_B + \cdots + f_{J\to M} Q_J$. Therefore, it results that

$$\frac{S_A^*}{\bar{a}_M^A \times t} = \frac{\varepsilon_M}{f_{A\to M}} = Q_A + \frac{f_{B\to M}}{f_{A\to M}} Q_B + \cdots + \frac{f_{J\to M}}{f_{A\to M}} Q_J\,. \tag{6.16}$$

The experimental expression $S_A^*/(\bar{a}_M^A \times t)$ then equals the rate of secretion of the tracee plus the sum of fractions of the rates of secretion of all other precursors. Each of these fractions may be smaller, equal to, or greater than unity, as they express the relative conversion of compounds B, C, ..., J to M, with respect to the conversion of A to the metabolite.

Two special cases of the general equation 6.16 have been already considered. When M is solely derived from A, then $S_A^*/(\bar{a}_M^A \times t) = Q_A$ (Case I) since all other terms in Eq. 6.16 will be zero, because either Q_X, $f_{X\to M}$, or both, are zero. When M is solely derived from secreted A and B (case II), it follows that

$$\frac{S_A^*}{\bar{a}_M^A \times t} = Q_A + \frac{f_{B\to M}}{f_{A\to M}} Q_B\,. \tag{6.17}$$

Conditions under which Eq. 6.16 represents the rate of production of the circulating hormone are discussed in another section of this chapter.

C. Parameters of Metabolism Estimated from Labeled Urinary Metabolites

Conversion of a Precursor to Urinary Metabolites

Perhaps the simplest type of tracer experiment designed to obtain quantitative information about metabolic pathways consists of measuring the extent of conversion of the administered tracer to one or more urinary metabolites (the f* fractions in the preceding section). Changes in metabolic patterns due to disease, to the influence of pharmacologic agents, or to normally variable physiologic conditions have been studied with tracers.

Relative Conversion of Two Precursors to a Metabolite

Another type of study involves the simultaneous administration of tracers of two precursors, labeled with different isotopes, and the measurement of the isotope ratio in a urinary metabolite. If, for instance, a mixture of tritiated compound A and [^{14}C]-labeled compound B is injected intravenously, the [^{3}H] : [^{14}C] ratio in a metabolite M indicates the relative extent to which the tracers were converted to the metabolite, i.e.,

$$\frac{\text{Fraction of } {}^{3}\text{H-A converted to M}}{\text{Fraction of } {}^{14}\text{C-B converted to M}} = \frac{[{}^{3}\text{H}] \text{ in M}/{}^{3}\text{H-A injected}}{[{}^{14}\text{C}] \text{ in M}/{}^{14}\text{C-B injected}} = \frac{({}^{3}\text{H}/{}^{14}\text{C}) \text{ in M}}{({}^{3}\text{H}/{}^{14}\text{C}) \text{ injected}}.$$

Experiments like this can be conveniently performed, since they involve measurements of isotope ratios rather than amounts of unlabeled compounds excreted in urine. Yet, this method is a powerful tool to study quantitative aspects of metabolism. KOPIN (1960) used the two-tracer technique to study the irreversible metabolism of epinephrine by analyzing urinary metabolites, VANDE WIELE *et al.* (1963) used it to study the reversible conversion of dehydroisoandrosterone to dehydroisoandrosterone sulfate, BAULIEU *et al.* (1965) used labeled androgens to study the origin of urinary androstanediol, etiocholanediol and other C_{19} steroids. SIITERI and MACDONALD (1963) studied the conversion of dehydroisoandrosterone sulfate (DS) to estrogens during pregnancy.

The transfer to peripheral circulation of an orally administered compound has also been estimated by two-tracer experiments. For instance, ^{14}C-DS was given orally whereas ^{3}H-DS was injected intravenously into the same subject (GURPIDE and VANDE WIELE, 1971). The isotope ratio in DS isolated from urine collected for an adequate period of time indicated the fraction of the ^{14}C-DS that entered the circulation. When pregnant women were studied, metabolites of DS formed extrasplanchnically (estrogens) served the same purpose. In fact, comparison of the administered isotope ratios to the ratios of an urinary metabolite of hepatic origin and an extraplanchnic metabolite provedes information about absorption of the compound from the digestive tract, hepatic clearance, as well as about transfer to circulation (TAIT and LITTLE, 1968).

Metabolites "Uniquely Derived" from a Circulating Compound

To calculate the amount of compound A converted to B per unit of time (e.g., in μmole/hr), specific activity data are necessary. For instance, to estimate the rate at which circulating A is excreted as M, it is necessary to determine the rate of *de novo* entry of A into systemic circulation (PR_A) as well as the fraction of A converted to M ($rate_{A \to M} = f_{A \to M} PR_A$).

Urinary methods to measure "production rates" and several other metabolic parameters are based on the concept of metabolites "uniquely derived" from a blood-borne compound. A metabolite M is defined as uniquely derived from a circulating compound A, if each molecule of the urinary metabolite originates from a molecule of A which has been in the systemic circulation.

If M and A are identical it can be assumed that M is uniquely derived from A, unless M were produced in the renal tubules from other precursors and excreted without entering the circulation.

Metabolite M can be uniquely derived from blood-borne A even if M has multiple secreted precursors. In this case, circulating A must be an obligatory intermediate in the conversion of all precursors to urinary M.

In contrast, M may not be uniquely derived from circulating A, even if compound A is a necessary chemical intermediate in the formation of M from all other precursors. This situation arises if A is formed in tissue from another precursor and is further metabolized before mixing with circulating A. Data indicating the existence of such a situation in the metabolism of testosterone have been presented by KORENMAN and LIPSETT (1964). They showed that urinary testosterone glucuronide, likely derived by glucuronidation of testosterone, is not uniquely derived from the circulating hormone; part of the testosterone formed in tissues from androstenedione was conjugated before being released to systemic circulation.

Relation Between the Specific Activities of a Circulating Compound and of a Urinary Metabolite Uniquely Derived from It

The rate at which compound A leaves the systemic circulation *irreversibly* and the rate at which the compound enters the circulation "de novo" (the production rate, PR_A) are equal at the steady state. In tracer experiments, the rate at which labeled A leaves the circulation irreversibly is $PR_A\, \sigma_A(t)$, where $\sigma_A(t)$, or simply σ_A, is the specific activity of blood-borne A. If the fraction f of this radioactivity appears in urine as metabolite M, the total amount of isotope to appear in that metabolite (E_M) is given by the expression

$$E_M = \int_0^\infty f \times PR_A \times \sigma_A \, dt = f \times PR_A \int_0^\infty \sigma_A \, dt .$$

If M is uniquely derived from circulating A, then the rate of urinary excretion of M (ε_M) equals $f \times PR_A$. Consequently,

$$E_M = \varepsilon_M \int_0^\infty \sigma_A \, dt . \tag{6.18}$$

Since, by definition, the cumulative specific activity of urinary M ($\bar{a}_M$) equals $E_M/(\varepsilon_M \times t)$, it then follows from Eq. 6.18 that

$$\bar{a}_M \times t = \int_0^\infty \sigma_A \, dt \,, \qquad (6.19)$$

if t, the time of urine collection, is large enough to allow for complete excretion of labeled M. As indicated by TAIT (1963), the urinary metabolite is providing, in this case, a physical integration of the specific activity of the circulating compound.

Measurement of Production Rates and Rates of Interconversion of Circulating Compounds

Equation 6.19, which is applicable to metabolites uniquely derived from a circulating compound, indicates that integral values of the specific activities of the blood-borne compound can be obtained from urinary data. Therefore, the formulas presented in Chapters 3 and 5 become relevant to the analysis of tracer experiments based on the isolation of urinary metabolites.

Production Rates

According to Eq. 3.2, the production rate of circulating A can be estimated by injecting a dose S_A^* of labeled A and measuring the specific activities of A in blood. Thus, $PR_A = S_A^* / \int_0^\infty \sigma_A \, dt$. Therefore, it follows that from Eq. 6.19

$$PR_A = \frac{S_A^*}{\bar{a}_M \times t} \qquad (6.20)$$

if M is uniquely derived from A and t is sufficiently large.

This convenient method can be applied when it is possible to isolate a metabolite M that is identical to A. In this case, the urinary metabolite can be reasonably assumed to be uniquely derived from A.

For instance, the production rate of blood-borne dehydroisoandrosterone sulfate has been measured from the specific activity of the compound in urine, in men and pregnant and nonpregnant women (SIITERI and MACDONALD, 1963; MACDONALD *et al.*, 1965). In one report (SANDBERG *et al.*, 1964), the production rate of this steroid was estimated simultaneously from blood and urinary data.

Since some unconjugated hormones are excreted in urine, cortisol among them (HSU and BLEDSOE, 1970), the urinary method may also be used to measure their rates of production.

It should be emphasized that if M is not uniquely derived from A, the expression $S_A^*/(\bar{a}_M \times t)$ represents not a production rate but rather a sum of terms as shown in Eq. 6.16. To call this expression the "urinary production rate" of the hormone is meaningless. Such a name, although in use by some authors, is not acceptable.

Rates of Interconversion

As shown in Chapter 5, the rates of interconversion between circulating compounds can be determined by measuring the areas under the specific activity curves of the compounds isolated from blood following the intravenous injection of tracers of these compounds. Thus, if a mixture of ^{3}H-A and ^{14}C-B is injected, the rate of conversion of circulating A to circulating B is given by the expression

$$v_{AB} = S_B^{14C} \frac{\int_0^\infty \sigma_B^{3H}\, dt}{\int_0^\infty \sigma_A^{3H}\, dt \int_0^\infty \sigma_B^{14C}\, dt - \int_0^\infty \sigma_B^{3H}\, dt \int_0^\infty \sigma_A^{14C}\, dt}. \tag{6.21}$$

If I and II metabolites uniquely derived from blood-borne A and B, respectively, it follows from Eqs. 6.19 and 6.21 that

$$v_{AB} = \frac{S_B^{14C}}{t} \frac{\bar{a}_{II}^{3H}}{\bar{a}_I^{3H}\, \bar{a}_{II}^{14C} - \bar{a}_{II}^{3H}\, \bar{a}_I^{14C}}. \tag{6.22}$$

The reader is referred to Table 3.1 in Chapter 3 for a description of the parameters that can be calculated from blood data following the injection of two tracers. All of these formulas can be applied to urinary data if metabolites uniquely derived from the circulating compounds are isolated.

D. Labeled Urinary Metabolites in Pregnancy

Procedures to estimate rates of secretion and metabolism of hormones during pregnancy using isotopic data obtained from blood-borne compounds are outlined in Chapter 5. Since adequate human fetal blood samples during pregnancy are not available, the design of tracer experiments using maternal urinary data is significant.

A distinction is again made in this section between the problem of estimating rates of secretion and rates of metabolism of the hormones. As before, the requirements to be imposed upon the origin of the urinary metabolites are less stringent when specific activities are used to estimate secretion rates rather than to calculate rates of metabolism.

Secretion Rates

A hormone can be produced during pregnancy by maternal and by fetal endocrine glands and also by the placenta, which may deliver it toward both the mother and the fetus at different rates. Single isotope experiments cannot be expected, in general, to yield values for any of these rates of secretion. Consider aldosterone as an example. To estimate the rate of secretion of this steroid in a non-pregnant subject, it appears sufficient to inject the labeled hormone intravenously and measure the specific activity of one of its urinary metabolites (Eq. 6.2). Basic assumptions to justify this calculation were that tracer and endogenously produced hormone had identical metabolic fate, and that the metabolite was not derived from another secreted precursor. During pregnancy, however, a tracer of aldosterone injected into a maternal vein cannot be assumed to have the same metabolic fate as the hormone produced by the fetal adrenals. In this case, Eq. 6.17 with A and B representing the maternal and fetal hormone, respectively, serves more adequately for the interpretation of the isotopic data. Thus, neither fetal or maternal rates of secretion nor the sum of these rates can be measured by this method. Note from Eq. 6.17, however, that if none of the fetal hormone were excreted as the urinary metabolite ($f_{B\to M} = 0$) or if the fetal hormone were not secreted ($Q_B = 0$) then the rate of secretion of the maternal hormone could be calculated. Since secretion may occur both to the maternal and fetal circulations, it is meaningful to refer to fetal and maternal rates of secretion of the placental hormone.

Administration of a single tracer into a maternal vein provides insufficient data to calculate rates of secretion of hormones in pregnancy. However, the rates Q_A and Q_B may be estimated if two tracers are administered. Examination of the procedure used for the derivation of Eqs. 6.12 and 6.13 reveals that these expressions are also applicable to the estimation of secretion rates of a hormone into maternal and fetal circulations. In this case, tracers A and B correspond to the same hormone labeled with a different isotope, e.g., ^{3}H-A and ^{14}C-B for steroids and ^{131}I-A and ^{125}I-B for protein hormones. One tracer is given to the mother and the other to the fetus. Metabolites I and II, used for the calculations, may originate by both fetal and maternal metabolism of the hormone.

In these experiments, the site of administration of the fetal tracer is important. The tracer should enter the fetal circulation and suffer a metabolic fate identical to that of the fetal endogenous hormone. As pointed out in Chapter 5, where methods based on the measurement of specific activities of hormones isolated from blood are considered, administration of the tracer into an umbilical vein may give different results than the administration of the same tracer into a fetal femoral vein (Gurpide *et al.*, 1972). In experiments in humans for which the urinary methods are most pertinent, direct administration of the tracers to the fetal circulation is generally not feasible. However, during blood transfusion *in utero*, a procedure performed in severe cases of erythroblastosis fetalis (Liley, 1963), it is possible to introduce the tracer with the transfusion blood. This situation has been exploited in a study on fetomaternal transfer and production of estradiol (Gurpide *et al.*, 1966 b).

Fetomaternal Transfers

Equations 6.21 and 6.22 can be applied to estimate the transfer of a hormone between the fetal and maternal circulations if metabolites "uniquely derived" from the hormone in each of these circulations were available. The closest approximation to this situation in human pregnancy has been the experiments involving administration of labeled estradiol into mother and fetus cited above. In this case, estradiol glucuronide in maternal urine was considered to be formed by the maternal organism from maternally circulating estradiol, while a labeled 15α-hydroxylated metabolite (estetrol) appeared to be formed by fetal metabolism of the estrogens in fetal circulation.

E. Comments

Tracer techniques to estimate rates of secretion and metabolism of hormones from data on specific activities of urinary metabolites are attractive because of their relative simplicity. However, interpretation of the data requires critical evaluation of the origin of the metabolites since they have multiple secreted precursors and are formed in several tissues. Advances in techniques for the measurement of hormones in blood have stimulated the use of circulating compounds as source of data in tracer experiments. However, labeled metabolites isolated from urine remain useful, particularly to estimate rates of production and interconversion of compounds which are very slowly metabolized, e.g., to study the conversion of dehydroisoandrosterone sulfate to 16α-hydroxydehydroisoandrosterone sulfate (Gurpide *et al.*, 1974). Production rates of circulating products, such as conjugated steroids which are cleared through the kidneys and appear in an identical form in blood and urine, can be conveniently estimated by urinary methods.

Furthermore, the current awareness of the intermittent character of the endocrine gland secretory processes may renew interest in the urinary metabolite methods to estimate average daily rates of production of the hormones.

Chapter 7

In vitro Tracer Superfusion Experiments

The preceding two chapters have dealt with the application of formulas derived from theoretical considerations discussed in Chapters 1 to 4 to *in vivo* experiments. This chapter deals with experiments performed *in vitro*, in which cell suspensions or tissue slices are superfused with medium containing tracers. These experiments yield steady-state isotopic data, which are interpreted on the basis of concepts presented in Chapters 1 and 2.

Superfusion Versus Batch Incubations

Tracer superfusion techniques show advantage over "batch" incubations in many respects. Measurement of rates at which hormones are formed or are metabolized, and elucidation of the relative importance of branched metabolic pathways are the subject of many investigations in which steroidogenic or target tissues are incubated "batchwise" with the labeled hormones. However, a recurring criticism of most of these experiments is that the results are dependent upon the incubation time. This pitfall is eliminated by the unrecycled superfusion of the tracers at a constant rate until an isotopic steady state is achieved. This technique also allows the study of steroid dynamics at near-physiologic levels of substrates. In contrast, batch incubations require either an excess of substrate to ensure excess supply of precursor or extremely short incubations to maintain the concentration of substrate practically unchanged. It is also of importance that superfusion of target tissue leads to complete replacement of the endogenous hormone by the exogenous tracer. Therefore, in nonsteroidogenic tissue, the intracellular concentration of hormones can be estimated from measurements of isotope content since the specific activity of each of the perfused tracers is known.

Superfusion experiments have other desirable characteristic which are unrelated to the use of tracers. A continuous flow of medium removes metabolic products that may alter enzymatic systems. Furthermore, continuous flow incubations of tissue fragments are particularly convenient to evaluate the effect of agents added to the medium on the production of hormones, as demonstrated by the studies on gonadal and adrenal steroidogenesis conducted by several investigators (TAIT *et al.*, 1967, 1970; SAFFRAN *et al.*, 1967; HUIBERGTSE and UNGAR, 1970).

Use of Two Metabolically Related Tracers in Superfusion Experiments

A fruitful experimental design (GURPIDE and WELCH, 1969) involves the exposure of tissue slices or cell suspensions, contained in a small chamber, to a continuous, unrecycled flow of a solution of two metabolically related compounds, one labeled with [^{3}H] and the other with [^{14}C]. Several fractions of perfusate are collected and, after an adequate time of superfusion, samples of tissue are removed from the chamber, washed thoroughly and homogenized. Concentrations of labeled and unlabeled compounds in perfusate (dpm/ml) and in tissue (dpm/g of tissue, dpm/mg of protein, or dpm/mg of DNA) are measured. From these experimental data, many relevant questions concerning the dynamics of production and metabolism of hormones can be studied.

One of the parameters that can be estimated with the superfusion technique is the fraction of a perfused tracer that enters the tissue. The determination of the unidirectional rate of entry of a compound into cells is required for the study of cell permeability and to investigate the existence and characteristics of saturable membrane transport systems for hormones. Another parameter that can be calculated from the same isotopic data describes the extent at which a hormone entering a cell is released back into the medium. The distinction between inward and outward flow of the material provides information that go beyond a measurement of net uptake, a parameter usually measured during batch incubations of tissue slices. Measurement of rates of release of hormones from cells to the medium, and the factors that can modify them, is of particular importance in the study of endocrine glands.

Not only rates of entry and exit can be calculated from isotopic data provided by the two-tracer perfusion systems; in addition, rates of interconversion between the two perfused compounds and rate at which these compounds are metabolized to other products can be estimated.

The simultaneous evaluation of all of these parameters yields a dynamic picture of intracellular processes at various concentrations of the hormones in the tissue. A comprehensive picture helps to interpret physiologic events occurring in target tissue and to search for agents that could modify hormonal action by altering the entry, metabolism, or release of the hormones. Similarly, the rates of release of particular steroids from endocrine glands is determined by a series of metabolic reactions, some of them reversible, occurring in the tissue. Measurement of these rates and evaluation of the factors that may influence them is a classic goal in endocrinology.

Model

An adequate representation of the parameters described above can be found in the unrestricted two pool model described by Figure 1.2, in Chapter 1. In this case, pools 1 and 2 represent two different compounds in the tissue. As described in Chapter 1, these two pools are considered to be embedded in a network of other exchanging compartments. Symbols v_{01} and v_{02} denote the rates of entry of the perfused compounds into the cells plus the rates of endogenous synthesis, if any. The rates v_{12} and v_{21} correspond to the intracellular interconversion of these compounds; v_{10} and v_{20} are the rates of removal of the compounds from the cells, either by metabolism or by exit to the medium. All these rates can be determined, as discussed in Chapter 2 and shown by the formulas in Table 2.1, if the rate of infusion of the tracers into pools 1 and 2 and the specific activities of the compounds in these pools at the steady state are known. In order to apply these calculations to in vitro superfusion systems, it is necessary to know what fraction of the superfused tracers actually enters the cell. Note that the rates I_1^* and I_2^* in Table 2.1 would not represent the rates of superfusion of the tracer but the actual rates at which the tracers enter the intracellular pools.

Calculation of Rates of Entry of Superfused Tracer into Cells

Consider the diagram in Fig. 7.1. Pool A represents compound A in the tissue. The superfused tracer of compound A may either enter the cells or bypass them. If α_A is the fraction of superfused ^{3}H-A entering the cells, $1-\alpha_A$ is the bypassing fraction. According to this definition,

$$1-\alpha_A = \frac{c_{\bar{A}_p}^{^3H} \text{ (bypassing tissue)}}{c_{\bar{A}}^{^3H}}\,. \tag{7.1}$$

The total concentration of ^{3}H-A in the perfusate, which is denoted by the symbol $c_{\bar{A}_p}^{^3H}$, is formed by tracer that bypassed the cells and ^{3}H-A, which, after entering the cells, is released back to the medium

$$c_{\bar{A}_p}^{^3H} = c_{\bar{A}_p}^{^3H} \text{ (bypassing tissue)} + c_{\bar{A}_p}^{^3H} \text{ (from tissue)}\,. \tag{7.2}$$

The estimation of α then requires an experimental distinction between the concentration in the perfusate of ^{3}H-A that either bypassed or entered and left the cells. This discrimination can be accomplished by intracellular labeling of compound A with another isotope. If ^{3}H-A and ^{14}C-B are superfused as a mixture and B is converted to A by enzymatic intracellular reactions, compound A in the tissue will be labeled with both [^{3}H] and [^{14}C]. At the isotopic steady state, the [^{3}H] : [^{14}C] ratio in intracellular compound A can be expected to be identical to the isotope ratio in compound A released from the cells into the medium. Therefore, by measuring the concentration of

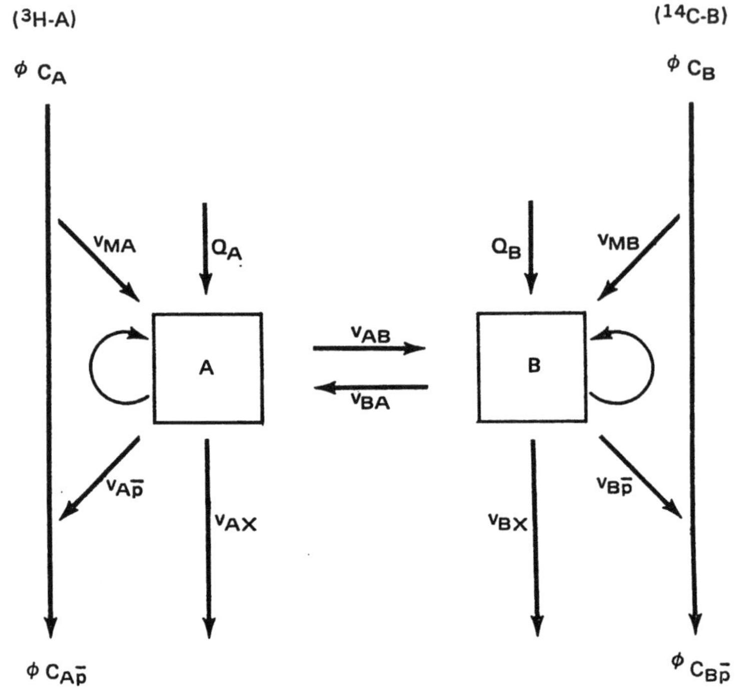

Fig. 7.1. Model representing two compounds, A and B, in tissue superfused with a mixture of ^{3}H-A and ^{14}C-B tracers, at a constant rate, without recycle. *Symbols:* Φ, flow rate (ml/hr); c_A and c_B, concentrations of A and B in the inflowing medium (nmole/ml); v_{MA} and v_{MB}, rates of entry of the compounds into cells (nmole/hr); v_{AB} and v_{BA}, rates of intracellular conversions between A and B; Q_A and Q_B, rates of synthesis of A and B from other intracellular precursors; $v_{A\bar{p}}$ and $v_{B\bar{p}}$, rates of exit of A and B from the tissue to the medium; v_{AX} and v_{BX}, rates of irreversible metabolism of A and B to all others products; $c_{A\bar{p}}$ and $c_{B\bar{p}}$, concentration of A and B in perfusate (outflowing medium). Reproduced from *Methods in Enzymology*, Vol. 36, O'Malley and Hardman, eds., Academic Press, New York

^{14}C-A in the perfusate ($c_{A\bar{p}}^{^{14}C}$), the concentration in the perfusate of ^{3}H-A derived from the tissue can be estimated. Thus, if

$$\frac{c_{A\bar{p}}^{^{3}H}\ \text{(from tissue)}}{c_{A\bar{p}}^{^{14}C}} = \left(\frac{^{3}H}{^{14}C}\right)_{AT} \tag{7.3}$$

then

$$c_{A\bar{p}}^{^{3}H}\ \text{(from tissue)} = \left(\frac{^{3}H}{^{14}C}\right)_{AT} c_{A\bar{p}}^{^{14}C} \tag{7.4}$$

where $(^{3}H/^{14}C)_{AT}$ denotes the isotopic ratio in compound A isolated from tissue at the steady state.

The formula to calculate α from the experimental data can be obtained from Eqs. 7.1 and 7.3:

$$\alpha_A = 1 - \frac{c_{A\bar{p}}^{^{3}H} - (^{3}H/^{14}C)_{BT}\, c_{A\bar{p}}^{^{14}C}}{c_B^{^{14}C}}\,. \tag{7.5}$$

Table 7.1. List of Formulas Applicable to the Model in Fig. 7.1 [a]

Fraction of superfused tracer entering cells

$$\alpha_A = 1 - \frac{c_{A\bar{p}}^{^3H} - (^3H/^{14}C)_{AT}\, c_{A\bar{p}}^{^{14}C}}{c_A^{^3H}} \qquad \alpha_B = 1 - \frac{c_{B\bar{p}}^{^{14}C} - (^{14}C/^3H)_{BT}\, c_{B\bar{p}}^{^3H}}{c_B^{^{14}C}}$$

Fraction of superfused tracer released by the tissue into the medium

$$\beta_A = (^3H/^{14}C)_{AT}\, c_{A\bar{p}}^{^{14}C}/c_A^{^3H} \qquad \beta_B = (^{14}C/^3H)_{BT}\, c_{B\bar{p}}^{^3H}/c_B^{^{14}C}$$

Conversion factors

$$\varrho_{AB} = \left(\frac{^3H}{^{14}C}\right)_{BT} \div \frac{\alpha_A}{\alpha_B}\left(\frac{^3H}{^{14}C}\right)_{\text{superfused}} \qquad \varrho_{BA} = \frac{\alpha_A}{\alpha_B}\left(\frac{^3H}{^{14}C}\right)_{\text{superfused}} \div \left(\frac{^3H}{^{14}C}\right)_{AT}$$

Rates

$$(D = a_{AT}^{^3H}\, a_{BT}^{^{14}C} - a_{BT}^{^3H}\, a_{AT}^{^{14}C})$$

$$v_{MA} + Q_A = \Phi\, \alpha_A\, c_A^{^3H}\, (a_{BT}^{^{14}C} - a_{AT}^{^{14}C})/D \qquad v_{MB} + Q_B = \Phi\, \alpha_B\, c_B^{^{14}C}\, (a_{AT}^{^3H} - a_{BT}^{^3H})/D$$

$$v_{MA} = \Phi\, \alpha_A\, c_A \qquad v_{MB} = \Phi\, \alpha_B\, c_B$$

$$v_{AB} = \Phi\, \alpha_B\, c_B^{^{14}C}\, a_{BT}^{^3H}/D \qquad v_{BA} = \Phi\, \alpha_A\, c_A^{^3H}\, a_{AT}^{^{14}C}/D$$

$$v_{A\bar{p}} = (\beta_A/\alpha_A)\, PR_A \qquad v_{B\bar{p}} = (\beta_B/\alpha_B)\, PR_B$$

$$PR_A = \Phi\, \alpha_A\, c_A^{^3H}/a_{AT}^{^3H} \qquad PR_B = \Phi\, \alpha_B\, c_B^{^{14}C}/a_{BT}^{^{14}C}$$

$$v_{A\bar{p}} + v_{AX} = \Phi\, (\alpha_A\, c_A^{^3H}\, a_{BT}^{^{14}C} - \alpha_B\, c_B^{^{14}C}\, a_B^{^3H})/D \qquad v_{B\bar{p}} + v_{BX} = \Phi\, (\alpha_B\, c_B^{^{14}C}\, a_{AT}^{^3H} - \alpha_A\, c_A^{^3H} a_{AT}^{^{14}C})/D$$

Intracellular clearances

$$IC_A = (PR_A/W) \div c_{AT} = (\Phi\, \alpha_A\, c_A^{^3H}/W) \div c_{AT}^{^3H}$$

$$IC_B = (PR_B/W) \div c_{BT} = \Phi\, \alpha_A\, c_B^{^{14}C}/W) \div c_{BT}^{^{14}C}$$

[a] Figure 7.1 illustrates the meaning of the symbols used to denote rates. The parameters α, β, and ϱ are defined as follows:

$$\alpha_A = \frac{v_{MA}}{\Phi\, c_A} \qquad \alpha_B = \frac{v_{MB}}{\Phi\, c_B}$$

$$\beta_A = \frac{v_{A\bar{p}}}{PR_A}\, \alpha_A \qquad \beta_B = \frac{v_{B\bar{p}}}{PR_B}\, \alpha_B$$

$$\varrho_{AB} = \frac{v_{AB}}{v_{AB} + v_{AX} + v_{A\bar{p}}} \quad \text{and} \quad \varrho_{BA} = \frac{v_{BA}}{v_{BA} + v_{BX} + v_{B\bar{p}}}$$

Φ is the rate of flow of the perfusion medium (ml/h), c_A and c_B are the concentrations (nmole/ml) of compounds A and B perfused. The symbol a_i^j denotes specific activity (dpm/nmole) of compound i with respect to isotope j. Subscripts T and $\bar{p}$ are used to indicate tissue and perfusate, respectively. The symbol [($^3H/^{14}C$) perfused] indicates the isotope ratio (dpm/dpm) in the perfusion medium, i.e., $c_A^{^3H}/c_B^{^{14}C}$.

If the conversion of B to A is reversible, as indicated in Fig. 7.1, the symmetry of the model makes it apparent that

$$\alpha_B = 1 - \frac{c^{^{14}C}_{A\bar{p}} - (^{14}C/^{3}H)_{BT}\, c^{^{3}H}_{A\bar{p}}}{c^{^{14}C}_{B}}\,.$$

Fraction of a Superfused Tracer Returning from the Cells to the Medium (β)

If β_A is defined as the ratio of the concentration in the perfusate of ^{3}H-A released by the cells and the concentration of the tracer in the perfusion medium, i.e.,

$$\beta_A = \frac{c^{^{3}H}_{A\bar{p}}\ (\text{from tissue})}{c^{^{3}H}_{A}}$$

it follows from Eq. 7.2 that

$$\beta_A = \frac{(^{3}H/^{14}C)_{AT}\, c^{^{14}C}_{A\bar{p}}}{c^{^{3}H}_{A}} \qquad (7.6)$$

Fraction of Superfused Tracer Appearing in the Perfusate as a Metabolite (γ)

The fraction of superfused ^{3}H-A appearing in the perfusate as a tritiated metabolite L is

$$\gamma_{AL} = \frac{c^{^{3}H}_{L\bar{p}}}{c^{^{3}H}_{A}}$$

Conversion Factors (ρ)

According to the definition of ϱ_{AB} given by Eq. 2.14 and the expressions indicating the entry of the tracer into the cells, it follows that

$$\varrho_{AB} = \frac{I^{^{14}C}_{B}}{I^{^{3}H}_{A}}\,\frac{a^{^{3}H}_{BT}}{a^{^{14}C}_{BT}} = \frac{(^{3}H/^{14}C)_{BT}}{(\alpha_A/\alpha_B)\,(^{3}H/^{14}C)\ \text{superfused}}\,. \qquad (7.7)$$

Similarly, $$\varrho_{BA} = \frac{I^{^{3}H}_{A}}{I^{^{14}C}_{B}}\,\frac{a^{^{14}C}_{AT}}{a^{^{3}H}_{AT}} = \frac{(\alpha_A/\alpha_B)\,(^{3}H/^{14}C)\ \text{superfused}}{(^{3}H/^{14}C)_{AT}}\,.$$

Rates in the Superfusion Model (Fig. 7.1)

Comparison of the models in Figs. 1.2 and 7.1 indicates that the rate of entry of material into pool A, denoted by v_{0A} in Fig. 1.2, appears resolved into two rates in Fig. 7.1, i.e.,

$$v_{0A} = v_{MA} + Q_A\,.$$

One of these rates (v_{MA}) corresponds to the entry of superfused compound A into the cells and is given by the expression

$$v_{MA} = \alpha_A \Phi c_A$$

where c_A is the concentration of A in the perfusion medium and Φ is the flow rate (e.g., ml/hr). The other rate (Q_A) represents the rate of intracellular formation of A from sources other than compound B. Since the total rate of entry into the pool (v_{0A}) can be calculated from the specific activities of A and B in the tissue (e.g., $a_{AT}^{^3H}$) and the concentrations of the labeled compounds in the perfusion medium and the perfusate, i.e.,

$$v_{0A} = \Phi \alpha_B c_B^{^{14}C} \frac{a_{AT}^{^3H}}{a_{AT}^{^3H} a_{BT}^{^{14}C} - a_{BT}^{^3H} a_{AT}^{^{14}C}}$$

then Q_A can be estimated by subtracting v_{MA} from v_{0A}.

A list of formulas for the calculation of rates from isotopic data obtained during superfusion experiments is given in Table 7.1.

The rate of removal of material from pool A represented by the rate v_{0A} in Fig. 1.2, appears in Fig. 7.1 resolved into two different rates: one ($v_{A\bar{p}}$) corresponds to the rate of release of A back into the medium, the other (v_{AX}) corresponds to the rate of irreversible metabolism of A to other products without the intermediacy of B. The estimation of the rate $v_{A\bar{p}}$ is based on the intuitively apparent relationship

$$\frac{v_{A\bar{p}}}{PR_A} = \frac{\beta_A}{\alpha_A}$$

which indicates that the ratio of the rate at which compound A is released by the tissue into the medium and the rate at which it appears *de novo* intracellularly equals the fraction of tracer A entering the tissue that is released back into the medium.

Then,
$$v_{A\bar{p}} = \frac{\beta_A}{\alpha_A} PR_A = \frac{\beta_A}{\alpha_A} \frac{\Phi \alpha_A c_A^{^3H}}{a_{AT}^{^3H}} = \Phi \frac{c_{A\bar{p}}^{^{14}C}}{a_{AT}^{^{14}C}}.$$

The rate of irreversible removal of A from the cells (v_{A0}, Fig. 1.2) can be estimated from the experimental data

$$v_{A0} = v_{A\bar{p}} + v_{AX} = \Phi \frac{\alpha_A c_A^{^3H} a_{BT}^{^{14}C} a_{BT} - \alpha_B c_B^{^{14}C} a_{BT}^{^3H}}{a_{AT}^{^3H} a_{BT}^{^{14}C} - a_{BT}^{^3H} a_{AT}^{^{14}C}}.$$

Consequently, v_{AX} can be calculated from the same data ($v_{AX} = v_{A0} - v_{A\bar{p}}$).

Intracellular Clearance (IC)

Another important parameter included in Table 7.1 is the intracellular clearance defined as the ratio of the production rate of the intracellular compound to the amount of compound in the tissue or, equivalently, as the ratio of the production rate per unit weight of tissue divided by the intracellular concentration of the compound, i.e.,

$$IC_A = \frac{PR_A}{W\, c_{AT}} = \frac{\alpha_A^{^3H} \Phi\, c_A^{^3H}}{W\, c_{AT}^{^3H}}$$

where W is the weight of the superfused tissue. This parameter has the dimensions of a rate constant (hr^{-1}). Other rate constants, corresponding to various processes of conversion or transfer of a compound, are obtained by dividing the rate corresponding to that process by the amount of the compound in the pool, e.g.,

$$k_{AB} = \frac{v_{AB}}{W\, c_{AT}}.$$

A Special Case: Nonsteroidogenic Tissue

When nonendocrine tissues or cell suspensions are used, the sole source of intracellular hormones at the steady state is the superfusion medium. Under these circumstances, the specific activity of compounds A and B in the tissue can be calculated knowing the amount of intracellular radioactivity in these compounds and the specific activity of the superfused tracers. Thus, if all intracellular A derives from superfused compounds A and B, it follows that at the steady state,

$$c_{AT} = \frac{c_{AT}^{^3H}}{a_A^{^3H}} + \frac{c_{AT}^{^{14}C}}{a_B^{^{14}C}}.$$

Concentrations of radioactivity are much more easily measured than specific activities, e.g., using isotope-dilution methods in which carriers of the compounds are added to the tissue homogenate or subcellular preparations.

Validation of the Model

Cell Viability

The survival and the metabolic stability of the superfused tissue can be evaluated by different means. The most direct is to perform measurements of oxygen consumption by determining the oxygen levels in the perfusate at various times during the superfusion. Several investigators have obtained practically constant steroidal output by pieces of superfused adrenal glands or slices during periods of 6 hours or longer. Furthermore, examination of the tissue by light or electron microscopy provides means to detect alterations occurring during superfusion. When cell suspensions are used, the technique of dye uptake may also be utilized to test for cell viability.

Isotopic Steady State

The achievement of a steady state in the concentrations of labeled compounds can be verified by measuring the levels of labeled compounds in perfusate and in tissue during superfusion. A more stringent test consists of measuring the concentrations of labeled hormones in nuclei isolated from superfused tissue at various times during the superfusion. The length of time required for the achievement of a steady state depends, of course, upon the rate constants of the processes in which the tracers are involved.

Compartmentalization

Although A appears as a single pool in the model in Fig. 7.1, it is not essential to consider that all of the intracellular compound is included in that pool. It should be noted that for many of the calculations outlined in this chapter isotope ratios and not concentrations of intracellular compounds are used (Eqs. 7.5 to 7.7). In these cases, the calculations are valid if the isotope ratios of a compound distributed in various compartments are identical at the steady state.

Compartmentalization can be investigated by comparing the specific activity of an intracellular compound, determined by direct analysis of the tissue, with a specific activity calculated indirectly. The indirect calculation of the specific activity of an intracellular compound from isotopic data is based on the following formula:

$$a_{AT}^{^3H} = \frac{c_{A_{\bar{D}}}^{^3H} - (1-\alpha_A)\, c_A^{^3H}}{c_{A_{\bar{D}}} - (1-\alpha_A)\, c_A}\,.$$

This expression indicates that the specific activity of the intracellular compound A is expected to equal the ratio of the rate at which labeled and unlabeled compound A are released from the tissue into the medium. Discrepancies of this calculated value with the experimentally determined specific activity could serve as evidence for the existence of nonmetabolic or passive pools of the compound in the tissue, as has been postulated for cholesterol. Furthermore, the exchangeable fraction of the total amount of the compound in the cell would equal the quotient of these two specific activities:

$$\text{Exchangeable fraction} = a_{AT}^{^3H} \div \left[\frac{c_{A_{\bar{D}}}^{^3H} - (1-\alpha_A)\, c_A^{^3H}}{c_{A_{\bar{D}}} - (1-\alpha_A)\, c_A}\right].$$

A powerful method to obtain evidence for intracellular compartmentalization of a compound involves perfusion of two of its precursors, labeled with different isotopes, and comparison of the isotopic ratio of the compounds in tissue and in perfusate at the steady state. Thus, if two precursors of X, labeled with different isotopes (e.g., ^{3}H-A and ^{14}C-B) are perfused as a mixture, and the isotope ratio in X in perfusate and tissue are found to be significantly different at the steady state, it can be concluded that more than one intracellular pool of X is present.

Applications

The original experimental design (GURPIDE and WELCH, 1969) and successive improved versions of the same (TSENG *et al.*, 1972) were developed to study steroid dynamics in human endometrium. Superfusion of slices of this tissue obtained at different stages of the menstrual cycle were used to determine whether the entry of estrogens and progesterone occurs by passive diffusion or by carrier-mediated mechanisms. For this purpose, the concentrations of the superfused steroids were kept constant in isotope content (dpm/ml) but were varied in a wide range in regard to mass (nmole/ml), resulting from the use of practically carrier-free tracers to saturated aqueous solutions. A kinetic expression characteristic of processes involving carrier mechanisms is as follows (STEIN, 1967):

$$\alpha_A = \frac{v_{MA}}{\Phi\, c_A} = \frac{v_{max}}{\Phi\,(K_M + c_A)}\,.$$

Findings indicating that the value of α_A remains constant during superfusions as various concentrations of A (c_A) can be interpreted to signify either that the apparent Michaelis constant (K_m) of interaction of the hormone with the carrier is much larger than the concentration of the hormone in the superfusion medium or that the entry occurs by diffusion. In the latter case, the rate of entry of the compound is proportional to its concentration in the medium and to the area available for entry (or to the weight of tissue if the geometry is maintained constant), i.e.,

$$\alpha_A = \frac{v_{MA}}{\Phi\, c_A} = \frac{k\,W}{\Phi}\,.$$

If the transfer of the steroid across the cell membrane is considered to occur by simple diffusion with similar permeability coefficience in both directions, it is possible to estimate the fraction of an intracellular compound, in a diffusible form (d) according to the following self-evident equation (GURPIDE and WELCH, 1969):

$$\frac{\beta_A}{\alpha_A} = \frac{d_A\, c_{AT}^{^3H}}{c_A^{^3H}} \quad \text{or} \quad d_A = \frac{\beta_A}{\alpha_A}\,\frac{c_A^{^3H}}{c_{AT}^{^3H}}\,.$$

The tracer superfusion method can be applied to the study of mechanisms by which different intermediates in biosynthetic reactions leave endocrine cells. The studies by superfusion of human endometrium with labeled estrone and estradiol made possible the measurements of rates of intracellular interconversion between these two compounds. These rates were measured at normal intracellular concentration and distribution of enzymes, cofactors and substrates. Since the intracellular concentration of substrates can be modified and measured, kinetic constants applicable to a reaction as it occurs in the cell can be determined.

Another important application of the tracer superfusion technique relates to the measurements of total hormone receptor concentration in target tissues. Prolonged superfusion of the tissue with labeled hormones of known specific activity results in a complete exchange of the intracellular hormone by the superfused labeled compound. The total receptor concentration can then be estimated by measuring the radioactivity associated with washed nuclei. This approach was used to study receptors of estradiol in human endometrium (TSENG and GURPIDE, 1972). Results obtained with batch incubations are usually affected by the presence of endogenous hormone competing with the tracer.

Application of the tracer superfusion system to the study of androgen dynamics in human prostate have been reported by GIORGI *et al.* (1971, 1973).

Studies on intracellular compartmentalization of uridine and uridine monophosphate in rat liver have also been published (TSENG and GURPIDE, 1973).

References

AGNEW, R. P.: Differential Equations. New York: MacGraw-Hill 1960.

ANDERSON, J., OSBORN, S. B., TOMLINSSON, R. W. S., WISE, M. E.: Clearance curves for radioactive tracers: Sum of exponential or powers of time? Phys. in Med. Biol. **14**, 498 (1969).

ARCOS, M., GURPIDE, E., VANDE WIELE, R. L., LIEBERMAN, S.: Precursors of urinary pregnanediol and their influence on the determination of the secretory rate of progesterone. J. clin. Endocr. **24**, 237 (1964).

BAIRD, D. T.: The secretion of androgens and oestrogens from the ovary and adrenal gland in reproductive endocrinology (W. J. Irvine, ed.). Edinburgh, London: E. S. Livingstone 1970.

BAIRD, D., HORTON, R., LONGCOPE, C., TAIT, J. F.: Steroid prehormones. Perspectives Biol. Med. Spring 384, 1968.

BAIRD, D. T., HORTON, R., LONGCOPE, C., TAIT, J. F.: Steroid dynamics under steady-state conditions. Recent Progr. Hormone Res. **25**, 611 (1969).

BALIKIAN, H. A., BRODIE, A. H., DALE, S. L., MELBY, J. C., TAIT, J. F.: Effect of posture on the metabolic clearance rate, plasma concentration and blood production rate of aldosterone in man. J. clin. Endocr. **28**, 1630 (1968).

BARDIN, C. W., LIPSETT, M. B.: Testosterone and androstenedione blood production rates in normal women and women with idiopathic hirsutism or polycystic ovaries. J. clin. Invest. **46**, 891 (1967).

BASSINGTHWAIGHTE, J. B.: Blood flow and diffusion through mammalian organs. Science **167**, 1347 (1970).

BASSINGTHWAIGHTE, J. B.: Circulatory transport and the convolution integral. Mayo Clin. Proc. **42**, 137 (1967).

BAULIEU, E. E., CARPÉCHOT, C., DRAY, F., EMILOZZI, R., LEBEAU, M. C., MAUVAIS JARVAIS, P., ROBEL, P.: An adrenal-secreted "androgen": Dehydroisoandrosterone sulfate. Its metabolism, and a tentative generalization on the metabolism of other steroid conjugates in man. Recent Progr. Hormone Res. **21**, 411 (1965).

BECKERS, C., CORNETTE, C.: TSH Production rate in nontoxic goiter. J. clin. Endocr. **32**, 852 (1971).

BEITINS, T. Z., KOWARSKI, A., SHERMETA, D. W., DE LEMOS, R. A., MIGEON, C. J.: Fetal and maternal secretion rate of cortisol in sheep: Diffusion resistence of the placenta. Pediatr. Res. **4**, 129 (1970).

BELLMAN, R.: "Dynamic Programming." Princeton: Princeton Univ. Press New Jersey 1957.

BERMAN, M., SCHOENFELD, R.: Invariants in experimental data on linear kinetics and the formulation of models. J. appl. Phys. **27**, 1361 (1956).

BERMAN, M., SHAN, E., WEISS, M. F.: The routine fitting of kinetic data to models. Biophys. J. **2**, 275 (1962).

BERSON, S. A., YALLOW, R. S.: Radioimmunoassay of ACTH in plasma. J. clin. Invest. **47**, 2725 (1968).

Bird, C. E., Green, R. N., Calandra, R. S., Connolly, J. G., Clark, A. F.: Kinetics of ^{3}H-testosterone metabolism in patients with carcinoma of the prostate: Effect of estrogen administration. Acta endocr. (Kbh) **67**, 733 (1971).

Box, G. E. P.: Fitting empirical data. Ann. N. Y. Acad. Sci. **86**, 792 (1960).

Bradley, E. M., Waterhouse, C.: Effect of estrogen administration on extravascular cortisol. J. clin. Endocr. **26**, 705 (1966).

Brown, J. B.: The relationship between urinary oestrogens and oestrogens produced in the body. J. Endocr. **16**, 202 (1957).

Brownell, G. L., Berman, M., Robertson, J. S.: Nomenclature for Tracer Kinetics. Int. J. appl. Radiat. **19**, 249 (1968).

Cameron, D. P., Burger, H. G., Catt, K. J., Doig, A.: Metabolic clearance rate of radioiodinated human growth hormone in man. J. clin. Invest. **48**, 1600 (1969).

Cavalieri, R., Searle, G. L.: The kinetics of distribution between plasma and liver of ^{131}I-labeled L-thyroxine in man. J. clin. Invest. **47**, 939 (1966).

Cavalieri, R. R., Steinberg, M., Searle, G. L.: Metabolic clearance rate of L-triiodothyronine in man: A comparison of results by single-injection and constant infusion methods. J. clin. Endocr. **33**, 624 (1971).

Chapdelaine, A.: Site of *in vivo* extraction and interconversion of testosterone and androstenedione in dogs. J. clin. Invest. **48**, 2063 (1969).

Chapdelaine, A., MacDonald, P. C., Gonzalez, O., Gurpide, E., Vande Wiele, R. L., Lieberman, S.: Studies on the secretion and interconversion of the androgens. IV. Quantitative results in a normal man whose gonadal and adrenal functions were altered experimentally. J. clin. Endocr. **25**, 1569 (1965).

Coble, Y. D., Kohler, P. O., Cargille, C. M., Ross, G. T.: Production rates and metabolic clearance rates of human follicle-stimulating hormone in premenopausal and postmenopausal women. J. clin. Invest. **48**, 359 (1969).

Cope, C. L., Black, E.: The production rate of cortisol in man. Brit. med. J. **1958** I, 1022.

Cox, D., Miller, H.: "The Theory of Stochastic Processes." New York: Wiley 1965.

Dazord, A., Saez, J., Bertrand, J.: Metabolic clearance rates and interconversion of cortisol and cortisone. J. clin. Endocr. **35**, 24 (1972).

Defares, J. G., Sneddon, I. N.: "The Mathematics of Medicine and Biology." Chicago, Illinois: Yearbook 1961.

de Lacerda, L., Kowarski, A., Migeon, C.: Diurnal Variation of the Metabolic Clearançe Rate of Cortisol. Effect on measurement of cortisol production rate. J. clin. Endocr. **36**, 1043 (1973).

Di Stefano, J. J. III, Furando, A. R., Jang, M., Jenkins, D., Johnson, D. J., Mak, P., Marshall, T., Mons, B., Warsavsky, A., Fisher, D. A.: Estimates and estimation errors of hormone secretion, transport and disposal rates in the maternal-fetal system. Endocr. **93**, 324 (1973).

Dixon, R., Human, A., Gurpide, E., Dyrenfurth, I., Cohen, H., Bowe, E., Engel, T., Daniel, S., James, S., Van de Wiele, R. L.: Feto-maternal transfer and production of cortisol in the sheep. Steroids **16**, 771 (1970).

Dominquez, R.: Kinetics of elimination, absorption and volume of distribution in the organism. In: Medical Physics (O. Glasser, ed.). Chicago, Illinois: Vol. II, Yearbook 1950.

Fabian, M., Lorsling, M. Z., Ibbitson, S., Jones, J. J., Stone, D.: The half-time ($t_{1/2}$) clearance, binding and stability of neurohypophysial hormones in human plasma. J. Physiol. (Lond.) **198**, 19 (1968).

FIELD, J. B.: Extraction by the liver. In: Handbook of Physiology, Section 7 (R. O. Greep and E. B. Astwood, eds.), Vol. I, American Physiological Society. Baltimore: Waverly Press 1972.

FLOOD, C., HUNTER, S. A., LLOYD, C. A., LONGCOPE, C.: The effect of posture on the metabolism of androstenedione and estrone in males. J. clin. Endocr. **36**, 1180 (1973).

FORD, H. C., PIETERS, H. P., BAILEY, R. E.: Aldosterone and sodium conservation: The effect of acute dietary sodium deprivation on the plasma concentration, the metabolic clearance and the secretion and excretion rates of aldosterone in normal subjects. J. clin. Endocr. **28**, 451 (1968).

GANT, N. F., HUTCHINSON, H. T., SIITERI, P. K., MACDONALD, P. C.: Study of the metabolic clearance rate of dehydroisoandrosterone sulfate in pregnancy. Amer. J. Obstet. Gynec. **111**, 555 (1971).

GIORGI, E. P., SHIRLEY, I. M., GRANT, J. K., STEWART, J. C.: Androgen dynamics *in vitro* in the human prostate gland. Effect of cyproterone and cyproterone acetate. Biochem. J. **132**, 465 (1973).

GIORGI, E. P., STEWART, J. C., GRANT, J. K., SCOTT, R.: Androgen dynamics *in vitro* in the normal and hyperplastic human prostate gland. Biochem. J. **123**, 41 (1971).

GURPIDE, E.: Mathematical analysis for the interpretation of *in vivo* tracer experiments. Acta endocr. (Kbh) Suppl. **158**, 26 (1972).

GURPIDE, E., MANN, J.: Estimation of secretory rates of hormones from the specific activities of metabolites which have multiple secreted precursors. Bull. Math. Biophys. **27**, 389 (1965).

GURPIDE, E., WELCH, M.: Dynamics of uptake of estrogens and androgens by human endometrium. J. biol. Chem. **244**, 5159 (1969).

GURPIDE, E., VANDE WIELE, R. L.: Utilization of ingested dehydroisoandrosterone sulfate for the production of estrogens by normal and adrenalectomized pregnant subjects. J. Steroid Biochem. **2**, 11 (1971).

GURPIDE, E., GIEBENHAIN, M., STOLEE, A., NOTATION, A., DIXON, R., BLACKARD, C. E.: Stimulation of 16α-hydroxylation of dehydroisoandrosterone sulfate by diethylstilbestrol. J. clin. Endocr. **37**, 867 (1973).

GURPIDE, E., MANN, J.: Interpretation of isotope data obtained from blood-borne compounds. J. clin. Endocr. **30**, 707 (1970).

GURPIDE, E., MACDONALD, P. G., VANDE WIELE, R. L., LIEBERMAN, S.: Measurement of the rates of secretion and of peripheral metabolism of two interconvertible compounds: Dehydroisoandrosterone-dehydroisoandrosterone sulfate. J. clin. Endocr. **23**, 346 (1963).

GURPIDE, E., MANN, J., LIEBERMAN, S.: Analysis of open systems of multiple pools by administration of tracers at a constant rate or as a single dose as illustrated by problems involving steroid hormones. J. clin. Endocr. **23**, 1155 (1963).

GURPIDE, E., MANN, J., SANDBERG, E.: Determination of kinetic parameters in a two-pool system by administration of one or more tracers. Biochemistry **3**, 1250 (1964).

GURPIDE, E., ROBERTS, K. D., WELCH, M. T., BANDY, L., LIEBERMAN, S.: Studies on the metabolism of blood-borne cholesterol sulfate. Biochemistry **5**, 3352 (1966 a).

GURPIDE, E., SCHWERS, J., WELCH, M. T., VANDE WIELE, R. L., LIEBERMAN, S.: Fetal and maternal metabolism of estradiol during pregnancy. J. clin. Endocr. **26**, 1355 (1966).

GURPIDE, E., TSENG, J., ESCARCENA, L., FAHNING, M., GIBSON, C., FEHR, P.: Fetomaternal production and transfer of progesterone and uridine in sheep. Amer. J. Obstet. Gynec. **113**, 21 (1972).

HAMILTON, W. F., MOORE, J. W., KINSMAN, J. M., SPURLING, R. G.: Studies on the circulation. IV. Further analysis of changes in hemodynamics under physiological and pathological conditions. Amer. J. Physiol. **99**, 534 (1932).

HART, H. E.: Analysis of tracer experiments: VII. General multicompartment systems imbedded in non-homogeneous inaccessible media. Bull. Math. Biophys. **28**, 261 (1966).

HEARON, J. Z.: Theorems on linear systems. Ann. N. Y. Acad. Sci. **108**, 36 (1963).

HEARON, J. Z.: Interpretation of tracer data. Biophys. J. **9**, 1363 (1969).

HELLMAN, L., NAKADA, F., CURTI, J., WEITZMAN, E. D., KREAM, J., ROFFWARG, H., ELLMAN, S., FUKUSHIMA, D. K., GALLAGHER, T. F.: Cortisol is secreted episodically by normal man. J. clin. Endocr. **30**, 411 (1970).

HEMBREE, W. C., BARDIN, C. W., LIPSETT, M. B.: A study of estrogen metabolic clearance rates and transfer factors. J. clin. Invest. **48**, 1809 (1969).

HIMMELBLAU, D. M.: Applied Nonlinear Programming. New York: McGraw-Hill 1972.

HORTON, R., TAIT, J. F.: Androstenedione production and interconversion rates measured in peripheral blood and studies on the possible site of its conversion to testosterone. J. clin. Invest. **45**, 301 (1966).

HORTON, R., TAIT, J. F.: *In vivo* conversion of dehydroisoandrosterone to plasma androstenedione and testosterone in man. J. clin. Endocr. **27**, 79 (1967).

HOWARD, C. W., ROBINSON, H., SCHMIDT, F. H., MCCORD, J. R., PREEDY, J. R. K.: Evidence for a two-pool system governing the excretion of radioactive urinary estrogen conjugates during the first eight hours following the administration of estrone-6,7-^{3}H to male subjects. J. clin. Endocr. **29**, 1618 (1969).

HSU, T. H., BLEDSOE, T.: Measurement of urinary free corticocoids by competitive protein-binding radioassay in hypoadrenal states. J. clin. Endocr. **30**, 443 (1970).

HUIBERGTSE, W. H., UNGAR, F.: Temporal and quantitative aspects of continuous superfusion of rat adrenals. Life sci. **9**, 349 (1970).

HUME, D. M.: The method of hypothalamic regulation of pituitary and adrenal secretion in response to trauma. In: Pathophysiologia Diencephalica (Curri, S. B., and Martini, L., eds.). Wien: Springer-Verlag 1958.

ITO, T., HORTON, R.: The source of plasma dihydrotestosterone in man. J. clin. Invest. **50**, 1621 (1971).

KAPLAN, S., GURPIDE, E., SCIARRA, J., GRUMBACH, M.: Metabolic clearance rate and production rate of chorionic growth hormone-prolactin in late pregnancy. J. clin. Endocr. **28**, 1450 (1968).

KARABOLAS, H. J., HUNTER, S. A., LLOYD, C. A., LONGCOPE, C.: Human placental 17β-estradiol dehydrogenase. J. biol. Chem. **244**, 4413 (1969).

KELLY, W. G.: Questions concerning the validity of one of the assumptions underlying the determination of the secretory rate of cortisol. Steroids **16**, 579 (1970).

KINSMAN, M. J., MOORE, J. W., HAMILTON, W. F.: Studies on the circulation. I. Injection method: Physical and mathematical considerations. Amer. J. Physiol. **89**, 322 (1929).

KOHLER, P. O., ROSS, G. T., ODELL, W. D.: Metabolic clearance and production rates of human luteinizing hormone in pre- and postmenopausal women. J. clin. Invest. **49**, 38 (1968).

KOPIN, I. J.: Technique for the study of alternate metabolic pathways; epinephrine metabolism in man. Science **131**, 1372 (1960).

KOPIN, I. J., AXELROD, J., GORDON, E.: Metabolic fate of H^3-epinephrine and C^{14}-metanephrine in the rat. J. biol. Chem. **236**, 2109 (1961).

KORENMAN, S. G., LIPSETT, M. B.: Is testosterone glucuronoside uniquely derived from plasma testosterone? J. clin. Invest. **43**, 2125 (1964).

KOUTRAS, D. A., BERMAN, M., SFONTOURIS, J., RIGOPOULOS, G. A., KOUKOULOMMATI, A. S., MALAMOS, B.: Endemic goiter in Greece: Thyroid hormone kinetics. J. clin. Endocr. **30**, 479 (1970).

KOWARSKI, A., LAWRENCE, B., HUNG, W., MIGEON, C. J.: Interconversion of cortisol and cortisone in man and its effects on the measurement of cortisol secretion rate. J. clin. Endocr. **29**, 377 (1969).

KOWARSKI, A., THOMPSON, R. G., MIGEON, C. J., BLIZZARD, R. M.: Determination of integrated plasma concentrations and true secretion rates or human growth hormone. J. clin. Endocr. **32**, 356 (1971).

KRIEGER, D. T., ALLEN, W., RIZZO, F., KRIEGER, H.: Characterization of the normal temporal pattern of plasma corticocoid levels. J. clin. Endocr. **32**, 266 (1971).

LARAGH, J. H., KELLY, W. G.: Aldosterone: Its biochemistry and physiology. In: Advanced Metabolic Disorders, vol. I (R. Levine and R. Luft, eds.). New York: Academic Press **1964**.

LARSON, K. B., SNYDER, D. L.: Measurement of relative blood flow, transit time distributions and transport-model parameters by residue detection when radiotracer recirculates. J. theor. Biol. **37**, 503 (1972).

LASSEN, N. A., LARSEN, O. A.: Measurement of blood flow with freely diffusible indicators as inert gases, antipyrine, labeled water rubidium. Acta endocr. (Kbh.) Suppl. **158**, 95 (1972).

LILEY, A. W.: Intrauterine transfusion of foetes in haemolytic disease. Brit. med. J. **1963 II**, 1107.

LIN, T. J., BILLIAR, R. B., LITTLE, B.: Metabolic clearance rate of progesterone in the menstrual cycle. J. clin. Endocr. **35**, 879 (1972).

LITTLE, B., BILLIAR, R. B.: Progesterone production. In: Progress in Endocrinology (C. Gual, ed.). Amsterdam: Exc. Med. Found. **184**, 270 (1969).

LITTLE, B., BILLIAR, R. B., HALLE, M., HEINSONS, A., JASSANI, M., KLINE, J. T., PURDY, R. H.: Pregnenolone production in pregnancy. Amer. J. Obstet. Gynec. **111**, 505 (1971).

LONGCOPE, C.: The metabolism of estrone sulfate in normal males. J. clin. Endocr. **34**, 113 (1972).

LONGCOPE, C., TAIT, J. F.: Validity of metabolic clearance and interconversion rates of estrone and 17β-estradiol in normal adults. J. clin. Endocr. **32**, 481 (1971).

LONGCOPE, C., KATO, T., HORTON, R.: Conversion of blood androgens to estrogens in normal adult men and women. J. clin. Invest. **48**, 2191 (1969).

LONGCOPE, C., LAYNE, D. S., TAIT, J. F.: Metabolic clearance rates and interconversions of estrone and 17β-estradiol in normal males and females. J. clin. Invest. **47**, 93 (1968).

LONGCOPE, C., WILLIAMS, K. I. H.: The metabolism of estrogens in normal women after pulse injections of ^{3}H-estradiol and ^{3}H-estrone. J. clin. Endocr. **38**, 602 (1974).

LORIAUX, D. L., RUDER, H. J., LIPSETT, M. B.: The measurement of estrone sulfate in plasma. Steroids **18**, 463 (1971).

LUSTCHER, J. A., COHN, A. P., CAMERGO, C. A., DOWDY, A. J., CALLAGHAN, A. M.: Aldosterone secretion and metabolism in hyperthyroidism and myxedema. J. clin. Endocr. **23**, 873 (1963).

MACDONALD, P. C., VAN DE WIELE, R. L., LIEBERMAN, S.: Precursors of the urinary 11-deoxy-17-ketosteroids: Influence of the level of secretion of dehydroisoandrosterone upon its metabolism. J. clin. Endocr. **22**, 1222 (1962).

MACDONALD, P. C., CHAPDELAINE, A., GONZALEZ, O., GURPIDE, E., VANDE WIELE, R. L., LIEBERMAN, S.: Studies on the secretion and interconversion of the androgens. III. Results obtained after the injection of several radioactive C_{19} steroids, singly or as mixtures. J. clin. Endocr. **25**, 1557 (1965).

MacDonald, P. C., Rombaut, R. P., Siiteri, P. K.: Plasma precursors of estrogens. J. clin. Endocr. **27**, 1103 (1967).

MacDonald, P. C., Grodin, J. M., Siiteri, P. K.: Dynamics of androgen and estrogen secretion. In: Steroid Secretion (D. Baird, ed.). Edinburgh, Scotland: Edinburgh Univ. Press 1971.

MacGillivray, M. H., Frohman, L. A., Doe, J.: Metabolic clearance and production rates of human growth hormone in subjects with normal and abnormal growth. J. clin. Endocr. **30**, 632 (1970).

Mahoudeau, J. A., Bardin, C. W., Lipsett, M. B.: The metabolic clearance rate and origin of plasma dihydrotestosterone in man and its conversion to the 5α-androstanediols. J. clin. Invest. **50**, 1338 (1971).

Mann, J., Gurpide, E.: Generalized rates of transfer in open systems of pools in the steady state. J. clin. Endocr. **26**, 1346 (1966).

Mann, J., Gurpide, E.: Interpretation of tracer data: Some factors which reduce the number of terms in the specific activity functions in n-pool systems. Bull. Math. Biophys. **31**, 473 (1969 a).

Mann, J., Gurpide, E.: Interpretation of tracer data: Significance of the number of terms in specific activity functions. Biophys. J. **9**, 810 (1969 b).

Marquardt, D. W.: An algorithm for least-squares estimation of nonlinear parameters. J. SIAM **11**, 431 (1963).

Meier, P., Zierler, K. L.: On the theory of the indicator dilution method for measurement of blood flow and volume. J. appl. Physiol. **6**, 731 (1954).

Midgley, A. R., Jaffe, R. B.: Regulation of human gonadotropins: X-episodic fluctuation of LH during the menstrual cycle. J. clin. Endocr. **33**, 962 (1971).

Migeon, C. J., Rivarola, M., Forest, M. G.: Studies of androgens in transexual subjects. Effects of estrogen therapy. Johns Hopkins Med. J. **123**, 128 (1968).

Nagai, N., Longcope, C.: Estradiol-17β and estrone: Studies on their binding to rabbit uterine cytosol and their concentration in plasma. Steroids **17**, 631 (1971).

Nahikian, H. M.: A modern algebra for biologists. Chicago, Illinois: Univ. of Chicago Press 1964.

Nicoloff, J. T., Low, J. C., Dussalt, J. H., Fisher, D. A.: Simultaneous measurement of thyroxine and triiodothyronine peripheral turnover kinetics in man. J. clin. Invest. **51**, 473 (1972).

Normand, M., Fortier, C.: Numerical versus analytical integration of hormonal disappearance data. Can. J. Physiol. Pharm. **48**, 274 (1970).

Odell, W. D., Utiger, R. D., Wilber, J. F., Condliffe, P. G.: Estimation of the secretion rate of thyrotropin in man. J. clin. Invest. **46**, 953 (1967).

Oppenheimer, J. H., Schwartz, A. L., Shapiro, H. C., Bernstein, G., Surks, M. I.: Differences in primary cellular factors influencing the metabolism and distribution of 3,5,3′-L-triidothyronine and L-thyroxine. J. clin. Invest. **49**, 1016 (1970).

Pearlman, W. H.: [16-^{3}H] progesterone metabolism in advanced pregnancy and in oophorectomized-hysterectomized women. Biochem. J. **67**, 1 (1957).

Pearlman, W. H., de Hertogh, R., Laumas, K. R., Pearlman, M. R. J.: Metabolism and tissue uptake of estrogen in women with advanced carcinoma of the breast. J. clin. Endocr. **29**, 707 (1969).

Rainville, E. D., Bedient, P. E.: Elementary Differential Equations. New York: MacMillan 1969.

Raus, J., Concolino, G., Gurpide, E., Lieberman, S.: Effect of preincubation of labeled testosterone with blood upon the metabolism of the tracer. In: Proc. of the Workshop Conference on Testosterone. Trembuttal (Tamm, J., ed.). Stuttgart: G. Thieme 1968, p. 109.

REBAR, R., PERLMAN, D., NAFTOLIN, F., YEN, S. S. C.: The estimation of pituitary luteinizing hormone secretion. J. clin. Endocr. **37**, 917 (1973).

RESCIGNO, A.: On transfer times in tracer experiments. J. theor. Biol. **39**, 9 (1973).

RESCIGNO, A., GURPIDE, E.: Estimation of average times of residence, recycle and interconversion of blood-borne compounds using tracer methods. J. clin. Endocr. **36**, 263 (1973).

RESIGNO, A., SEGRE, G.: Drug and Tracer Kinetics. Waltham, Massachusetts: Blaisdell 1966.

RIGGS, D. S.: The Mathematical Approach to Physiological Problems. Baltimore, Maryland: Williams & Wilkins 1963.

RIVAROLA, M. A., SAEZ, J. M., MEYER, W. J., JENKINS, M. E., MIGEON, C. J.: Metabolic clearance rate and blood production rate of testosterone and androstenedione under basal conditions, ACTH and HCG stimulation, comparison with urinary production rate of testosterone. J. clin. Endocr. **26**, 1208 (1966).

RIVAROLA, M. A., SINGLETON, R. T., MIGEON, C.: Splanchnic extraction and interconversion of testosterone and androstenedione in man. J. clin. Invest. **46**, 2095 (1967 a).

RIVAROLA, M. A., SAEZ, J. M., MEYER, W. J., KENNY, F. M., MIGEON, C. J.: Studies of androgens in the syndrome of male pseudohermaphroditism with testicular feminization. J. clin. Endocr. **27**, 371 (1967 b).

RIZKALLAH, T., GURPIDE, E., VANDE WIELE, R. L.: Metabolism of human chorionic gonadotropin in man. J. clin. Endocr. **29**, 92 (1969).

ROBERTSON, J. S.: Theory and use of tracers in determining transfer rates in biological systems. Physiol. Rev. **37**, 133 (1957).

ROMANOFF, L. P., MORRIS, C. W., WELCH, P., GRACE, M. P., PINCUS, G.: Metabolism of progesterone-4-C^{14} in young and elderly men. J. clin. Endocr. **23**, 286 (1963).

ROTHSCHILD, M. A., BAUMAN, A., YALLOW, R. S., BERSON, S. A.: The effect of large doses of desiccated thyroid on the distribution and metabolism of albumin-I^{131} in euthyroid subjects. J. clin. Invest. **3**, 422 (1957).

RUDER, H. J., LORIAUX, D. L., LIPSETT, M. B.: Estrone sulfate: Production rate and metabolism in man. J. clin. Invest. **51**, 1020 (1972).

RUDOLPH, A. M., HEYMANN, M. A.: Measurement of flow in perfused organs using microsphere techniques. Acta endocr. (Kbh.) Suppl. **158**, 112 (1971).

SAEZ, J. M., BERTRAND, J., MIGEON, C. J.: Metabolic clearance rate, urinary and plasma production rates of testosterone sulfate in man. Steroids **17**, 435 (1971).

SAEZ, J. M., FOREST, M. G., MORERA, A. M., BERTRAND, J.: Metabolic clearance rate and blood production rate of testosterone and dihydrotestosterone in normal subjects during pregnancy and in hyperthyroidism. J. clin. Invest. **51**, 1226 (1972).

SAFFRAN, M., FORD, P., MATTHEWS, E. K., KRAML, M., GARBACZEWSKA, L.: An automated method for following the production of corticosteroids by adrenal tissue *in vitro*. Canad. J. Biochem. **45**, 1901 (1967).

SANDBERG, A. A., ROSENTHAL, H., SCHNEIDER, S. L., SLAUNWHITE, N. R., JR.: Protein—steroid interactions and their role in the transport and metabolism of steroids. In: Steroid Dynamics (G. Pinkus, T. Nakoa, J. F. Tait, eds.). New York: Academic Press 1966.

SANDBERG, E., GURPIDE, E., LIEBERMAN, S.: Quantitative studies on the metabolism of dehydroisoandrosterone sulfate. Biochemistry **3**, 1256 (1964).

SCHAMBELAN, M., BIGLIERI, E. G.: Deoxycorticosterone production and regulation in man. J. clin. Endocr. **31**, 695 (1972).

SEGRE, E. J., FRIEDRICH, E. H., DODEK, O. E., LLOYD, C. W., LOBOTSKY, J., LEVIN, J., KLEIBER, E. L.: Effects of epinephrine on the production and metabolic clearance of cortisol in normal men and women and in women with idiopathic hirsutism. Acta endocr. (Kbh.) **53,** 561 (1966).

SELBY, S. M., ed.: Handbook of Tables for Mathematics. Cleveland, Ohio: CRC Press 1972.

SHARNEY, L., WASSERMAN, L. R., GEVIRTZ, N. R.: Representation of certain mammillary n-pool systems by two pool models. Med. Electron. **3,** 249 (1964).

SHIPLEY, R. A., CLARK, R. E.: Tracer Methods for *In Vivo* Kinetics. Theory and application. New York: Academic Press 1972.

SIITERI, P. K., MACDONALD, P. C.: The utilization of circulating dehydroisoandrosterone sulfate for estrogen synthesis during human pregnancy. Steroids **2,** 713 (1963).

SILVERS, A., SWENSON, R. S., FARQUHAR, J. W., REAVEN, G. M.: Derivation of a three compartment model describing disappearance of plasma insulin-^{131}I in man. J. clin. Invest. **48,** 1461 (1969).

SINGER, P. A., NICOLOFF, J. T.: Metabolic clearance rates of diiodotyrosine in man. J. clin. Endocr. **34,** 666 (1972).

SKINNER, S. M., CLARK, R. E., BAKER, N., SHIPLEY, R. A.: Complete solution of the three-compartmental model in steady-state after single injection of radioactive tracer. Amer. J. Physiol. **196,** 238 (1959).

SOLOMON, A. K.: Compartment Methods of Kinetic Analysis in Mineral Metabolism (C. L. Comer and F. Bronner, eds.). New York: Academic Press 1960.

SOUTHREN, A. L., TOCHINOTO, S., CARMODY, N. C., ISURGI, K.: Plasma production rates of testosterone in normal adult men and women and in patients with the syndrome of feminizing testes. J. clin. Endocr. **25,** 1441 (1965).

SOUTHREN, A. L., GORDON, G. C., TOCHINOTO, S.: Further study of factors affecting the metabolic clearance rate of testosterone in man. J. clin. Endocr. **28,** 1105 (1968).

STEELE, R.: Tracer Probes in Steady State Systems. Springfield, Illinois: Thomas 1971.

STEIN, W. D.: The Movement of Molecules Across Cell Membranes. New York: Academic Press 1967.

STEWART, G. N.: Research on the circulation time and on the influence which affect it. IV: The output of the heart. J. Physiol. (Lond.) **22,** 159 (1897).

STROTT, C. A., BERMUDEZ, J. A., LIPSETT, M. B.: Blood levels and production rate of 17-hydroxypregnenolone in man. J. clin. Invest. **49,** 1999 (1970).

SURKS, M. I., SCHADLOW, A. R., STOCK, J. M., OPPENHEIMER, J. H.: Determination of iodothyronine absorption and conversion of L-thyroxine (T_4) to L-triiodothyronine (T_3) using turnover rate techniques. J. clin. Invest. **52,** 805 (1973).

TAIT, J. F.: Review: The use of isotopic steroids for the measurement of production rates *in vivo*. J. clin. Endocr. **23,** 1285 (1963).

TAIT, J. F., BURSTEIN, S.: *In vivo* studies of steroid dynamics in man. In: The Hormones (G. Pincus, K. V. Thinmann, E. B. Astwood, eds.). New York: Academic Press 1964.

TAIT, J. F., LITTLE, B.: The metabolism of orally and intravenously administered labeled aldosterone in pregnant subjects. J. clin. Invest. **47,** 2423 (1968).

TAIT, J. F., TAIT, S. A. S., LITTLE, B., LAUMAS, K. R.: The disappearance of 7-^{3}H-D-aldosterone in the plasma of normal subjects. J. clin. Invest. **40,** 72 (1961).

TAIT, J. F., LITTLE, B., TAIT, S. A. S., FLOOD, C.: The metabolic clearance rate of aldosterone in pregnant and non-pregnant subjects estimated by both single injection and constant infusion method. J. clin. Invest. **41,** 2093 (1962).

TAIT, S. A. S., TAIT, J. F., FLOOD, C., OKAMOTO, M.: Production of steroids by *in vitro* superfusion of endocrine tissue. I. Apparatus and a suitable analytical method for adrenal output. Endocrinology **81,** 1213 (1967).

Tait, S. A. S., Schulster, D., Okamoto, M., Flood, C., Tait, J. F.: Production of steroids by *in vitro* superfusion of endocrine tissue. II. Steroid output from bisected whole and decapsulated adrenals of normal intact, hypophysectomized and hypophysectomized nephrectomized rats as a function of time of superfusion. Endocrinology **86**, 360 (1970).

Taylor, A. L., Finster, J. L., Mintz, D. H.: Metabolic clearance and production rates of human growth hormone. J. clin. Invest. **48**, 2349 (1969).

Thompson, R. G., Rodriquez, A., Kowarski, A., Blizzard, R. M.: Growth hormone: Metabolic clearance rates, integrated concentration, and production rates in normal adults and the effect of prednisone. J. clin. Invest. **51**, 3193 (1972).

Thron, C. D.: Structure and kinetic behavior of linear Multicompartmental systems. Bull. Math. Biophys. **34**, 277 (1972).

Tseng, J. K., Gurpide, E.: Compartmentalization of uridine and uridine 5′-monophosphate in rat liver slices. J. biol. Chem. **248**, 5634 (1973).

Tseng, L., Gurpide, E.: Nuclear concentration of estradiol in superfused slices of human endometrium. Amer. J. Obstet. Gynec. **114**, 995 (1972).

Tseng, L., Stolee, A., Gurpide, E.: Quantitative studies on the uptake and metabolism of estrogens and progesterone by human endometrium. Endocrinology **90**, 390 (1972).

Turner, R. C., Grayburn, J. A., Newman, G. B., Nabarro, J. D. N.: Measurement of the insulin delivery rate in man. J. clin. Endocr. **33**, 279 (1971).

Ulick, S., Laragh, J. H., Lieberman, S.: The isolation of a urinary metabolite of aldosterone and its use to measure the rate of secretion of Aldosterone by the adrenal cortex of man. Trans. Ass. Amer. Phycns **71**, 225 (1958).

Vande Wiele, R. L., MacDonald, P. C., Gurpide, E., Lieberman, S.: Studies on the secretion and interconversion of the androgens. Recent Progr. Hormone Res. **19**, 275 (1963).

Wade, T. L.: The Algebra of Vectors and Matrices. Reading, Massachusetts: Addison-Wesley 1951.

Wagner, J. G.: Pharmacokinetics. Ann. Rev. Pharmacol. **8**, 67 (1968).

Wang, D. Y., Bulbrook, R. D., Ellis, F., Comb, M. M.: Metabolic clearance rates of pregnenolone, 17-acetoxypregnenolone and their sulfate esters in man and in rabbit. J. Endocr. **39**, 395 (1967 a).

Wang, D. Y., Bulbrook, R. D., Sneddon, A., Hamilton, T.: The metabolic clearance rate of dehydroisoandrosterone, testosterone and their sulfate esters in man, rat and rabbit. J. Endocr. **38**, 307 (1967 b).

Waterhouse, C., Keilson, J.: Transfer times across the human body. Bull. Math. Biophys. **34**, 33 (1972).

Whittaker, E., Robinson, G.: The Calculus of Observations. London: Blackie & Son, Ltd. 1924 (reprinted, Dover, N. Y., 1967).

Woeber, K. A., Sobel, R. J., Ingbar, S. H., Sterling, K.: The peripheral metabolism of triiodothyronine in normal subjects and in patients with hyperthyroidism. J. clin. Invest. **49**, 643 (1970).

Yen, S. S. C., Tsai, C. C., Naftolin, F., Vandenberg, G., Ajabor, L.: Pulsatile patterns of gonadotropin release in subjects with and without ovarian function. J. clin. Endocr. **34**, 671 (1972).

Zierler, K. L.: Effect of circulatory beds on tracer experiments, or noncompartmental analysis. Ann. N. Y. Acad. Sci. **108**, 106 (1963).

Zilversmit, D. B., Entenman, C., Fishler, M. C.: On the calculation of "turnover time" and "turnover rate" from experiments involving the use of labeling agents. J. gen. Physiol. **26**, 325 (1943).

Subject Index

MIX
Papier aus verantwortungsvollen Quellen
Paper from responsible sources
FSC® C105338

If you have any concerns about our products,
you can contact us on
ProductSafety@springernature.com

In case Publisher is established outside the EU,
the EU authorized representative is:
Springer Nature Customer Service Center GmbH
Europaplatz 3, 69115 Heidelberg, Germany

Printed by Libri Plureos GmbH
in Hamburg, Germany